A Study of
Inner Space

Oceanography

Warren E. Yasso

Holt, Rinehart and Winston, Inc., New York

Dr. Warren E. Yasso is Associate Professor, Department of Science Education, Teachers College, Columbia University, New York. Since 1958, with support from U. S. Navy, Office of Naval Research, Geography Branch, Dr. Yasso has been investigating problems of beach morphology and shallow-water oceanography at Sandy Hook, New Jersey. He received his Ph.D. in Geology from Columbia University in 1964.

4567890 076 98

Printed in United States of America

ISBN 0-03-044735-6

Preface

Oceanography in the United States educational system is presently considered to be an advanced program of graduate study for those well grounded in one of the basic sciences or engineering. But the secondary school student, the undergraduate college student, and the interested layman need not think themselves excluded from the fraternity of oceanographers. One of the best illustrations of the important role the interested layman can play in oceanographic research is the annual Labor Day fish count made by amateur SCUBA divers, members of the American Littoral Society, working under direction of the Department of the Interior, Fish and Wildlife Service. By tallying number and kind of fish observed in small units of shallow coastal water the amateur diver performs an oceanographic research function which, when combined with data from other divers, represents essential information on certain fish populations. Many other projects like this will arise in oceanography. This fledgling science needs the moral and financial support of every citizen, personally, as well as through elected representatives. With cooperation of its citizenry the United States may realize the economic and social benefits that await the first nation to truly conquer the oceans.

Throughout this book the Metric System of measurement units has been given primary emphasis. It would seem that with the impetus supplied by international scientific cooperation such as I.G.Y. (International Geophysical Year) and I.Q.S.Y. (International Quiet Sun Year) metric units have become so commonplace that students destined for a scientific career will soon use them to the exclusion of other systems. To ease the burden of conversion I have indicated an equivalent English System unit for each measurement given.

Within the space limitations imposed I have attempted to describe only the potential and progress of deep-sea oceanography. In so doing almost the entire subject matter of nearshore oceanography has been passed over. Such fascinating topics as waves and tides, currents, beaches and coast lines, and estuarine processes will have to await exposition in some future book of this series. Also, it was decided not to include a detailed discussion of continental drift or crust and mantle interaction because these topics would require presentation of extensive factual background for the reader to use in forming a fair judgment of the alternative hypotheses available. Similarly, results of investigating the thermal, sonic, magnetic, and gravity characteristics of the oceanic crust have been passed over until such time as they can be used in conjunction with a more incisive analysis of ocean basin formation and development.

I have attempted to extract significant facts from the standard and current oceanographic literature without, in most cases, giving the source reference. I trust that the unacknowledged use of the research efforts of others may be understood within the context of a book designed to be merely an elementary introduction to marine science.

Photographs and descriptive material supplied by agencies of the United States government and by non-governmental marine laboratories and their personnel are gratefully acknowledged. Many U.S. oil companies and manufacturers of oceanographic "hardware" supplied photographs and technical data which add greatly to this presentation. Especially, I thank Professor Bruce C. Heezen, Lamont Geological Observatory, Columbia University, whose gifted teaching supplied much of the background on which this book is built. However, Dr. Heezen is in no way responsible for any errors of fact or interpretation that are contained herein. To my wife, Marian, our family, relatives, friends, and the editorial staff of Holt, Rinehart and Winston I express my appreciation for their patience, understanding, and inspiration during preparation of this manuscript.

W.E.Y.

Contents

1

Introduction

To many of us, certain facts about the earth's oceans have become so familiar—have been repeated so often—that they are almost clichés: Oceans cover nearly three-fourths of the earth's surface; the topography of the ocean floors is more rugged and more mountainous than the topography of the continents; the greatest oceanic depths are greater than the height of Mt. Everest. Indeed, these facts are clichés in a sense, for they overemphasize what is almost commonplace, and obscure what is truly remarkable.

Consider, for example, the fact that the vertical distance from the top of Mt. Everest to the bottom of the Mariana Trench in the west Pacific is about 20 kilometers, or slightly less than 12½ miles. Even if such height differences were the rule rather than the exception, the solid surface of the earth would still be—relatively speaking—much smoother than the surface of an orange. But the oceans are even "smoother" than that. The volume of the oceans is roughly 1365 million cubic kilometers (325 million cubic miles), with a mass of 1560 million billion tons. Although these are enormous figures, they represent only *eight ten-thousandths of the volume* and *240 millionths of the mass* of the earth. Covering a surface area of 362 million square kilometers (141 million square miles), the average depth of this water mass is slightly less than 3.8 kilometers, or about 2.35 miles. Truly, the oceans are hardly more than a film of salt water on the surface of our planet!

Yet, it was in some shallow, near-shore area of this film that life on earth began two or three billion years ago. Today it is the tremendous abundance of life in the sea that appears to present the surest solution of the world's food problems. Well over half—perhaps as much as 85 percent—of the food product of all plants is produced by marine plants. In time the oceans will certainly become a major source of valuable minerals and chemicals and an important source of fresh water when these are no longer available in sufficient quantities on the continents. Also, in recent years, it has become clear that the oceans play a very important but little-understood role in determining the earth's weather and climate patterns. And finally, it may be that the sediment layers at the bottom of the sea contain a record going back several billion years—a record of the earth's history. For these and many other reasons, the importance of the earth's oceans can hardly be overestimated.

Nonetheless, the importance of the oceans has become widely recognized only in recent years. It is surprising that the study of the largest single feature of the earth's surface today involves the services of a few thousand scientists and technicians; for almost 61 percent of the Northern Hemisphere and about 81 percent of the Southern Hemisphere is covered by water. Yet the exploration of this "inner space" lags behind the exploration of outer space. The United States National Oceanographic Data Center reports that slightly over 300 research vessels, representing 32 nations, were in active service during 1963. But even this small number of vessels represents a phenomenal increase in ships and personnel since the rebirth of oceanography at the end of World War II.

Oceanography is a relatively young science. For purposes of simplification, it may be defined as the *study of the oceans.* Oceanographers are drawn from the four large areas of science: geology, chemistry, physics, and biology. Geologic study of oceanic sediments, rock structure, and topography is referred to as *submarine geology* or *geological oceanography.* The study of sea-water chemistry is called *marine chemistry* or *chemical oceanography.* The biological aspects of the ocean environment, such as fish populations and plant life, are studied by the *marine biologist,* or *biological oceanographer.* Atmospheric processes, circulation of the oceans, and the physical properties of ocean water fall within the scope of *marine physics,* or *physical oceanography.*

In the past the study of the oceans was largely a search for new trade routes. Before we consider the various areas of oceanographic research, therefore, it will be worthwhile to review very briefly the historical background of this science.

The Historical Background

A discussion of the history of oceanography should begin with the first human to recognize and solve a problem associated with the oceans—the pioneer oceanographer. Perhaps he sought a greater catch of fish or perhaps he sought to reach an island visible across a short stretch of open water. Whatever his reason for exploring the oceans, his name was forgotten long before mankind learned how to record facts of historical importance. But there were many men who explored the oceans whose names are familiar to us and whose search for trade and conquest resulted in much practical knowledge about the geographical relationships between continents and ocean basins. The political and economic ties that developed from these early European voyages of discovery are still important today.

Captain Bartholomeu Diaz rounded the Cape of Good Hope in 1488, but it was not until 1497–98 that Vasco da Gama sailed beyond this point in the hope of finding a sea route to India—a mission much like that of Christopher Columbus. When Admiral da Gama rounded the Cape of Good Hope in April, 1498, he acquired the services of an Arab navigator, Ahmad ibn Majid, to guide him across the Indian Ocean to the Malabar Coast of India. Although the route was not known to European navigators, the Arabs had an intimate knowledge of sea routes between the Indian Ocean, Arabian Sea, and Mediterranean Sea. It is said that the name Gibraltar comes from the name "Jebel Tariq" (meaning "Mount of Tariq") given it by these early Arabic navigators.

In 1519, Ferdinand Magellan sailed from Spain with five ships, on a mission to find a passage between the Atlantic and Pacific Oceans. His route was between the South American mainland and Tierra del Fuego, by way of what we now call the Straits of Magellan. Merchant ships used this route until Sir Francis Drake found a more practical route south of Cape Horn.

Captain James Cook, between 1769 and 1779, commanded British vessels on three major voyages of discovery. On the first of these, he discovered Australia; on the last Cook set off in search of a northwest passage between the Atlantic and Pacific Oceans and successfully penetrated into the Arctic Ocean by way of the Bering Strait. A short distance beyond the Bering Strait, he was stopped by the Arctic ice pack. Returning to the Hawaiian Islands (which he had rediscovered in February, 1776), Cook was killed in a battle with natives.

As the volume of ocean traffic increased after the great voyages of discovery, much practical knowledge about the speed and direction of ocean currents and winds was stored in the minds and log books of ships' officers. One of the most impressive compilations of such oceanographic information was published in 1855 by Lieutenant Matthew Fontaine Maury of the United States Navy.

Maury began his naval service in 1825 as a midshipman; no U.S. Naval Academy existed at the time and so his entire training took place aboard ship. In 1834 an injury caused him to be assigned to land duty. He immediately began to publish articles on navigation. With his appointment as Superintendent of the Navy Depot of Charts and Instruments in 1842, Maury was able to examine the logbooks kept by many captains. A combination of these log-book studies and his own practical experience led to the publication of the first book on oceanography to be seen by the English-speaking world, Maury's *The Physical Geography of the Sea*. The book, an immediate "best-seller" both for navigators and laymen, remains of great historical significance.

The Beginnings of Deep-sea Research

Even though man came to know more about the ocean surface, beneath it lay a dark, unknown, forbidding world. Man could not go far below the surface to see for himself what this alien environment was like. He had to invent devices–sensors –which would tell him things about the depth of the oceans that he could not experience directly through his own senses of sight, sound, and touch. The first depth measurements were made in the Sea of Sardinia early in the second century B.C. The first known attempt at sounding the deep ocean was made by Magellan in 1521 during his attempt to circumnavigate the globe. Between 1839 and 1843 Captain James C. Ross led two ships on an expedition to plumb the ocean depths and chart the Antarctic Ocean. The first successful deep-sea soundings were made during this voyage.

Modern oceanographers consider scientific exploration of the oceans to have begun with the *Challenger* expedition of 1873–1876. *H.M.S. Challenger* was an 18-gun corvette of the British Navy that had been stripped of battle gear and fitted out for oceanographic research. It had a displacement of 2300 tons and was equipped with both sail and steam power. The British Admiralty provided a crew under the command of Captain George Nares; Charles Wyville Thomson, Professor of Natural Philosophy at the University of Edinburgh, headed the research staff. The 240-man expedition left Portsmouth, England, in December, 1872. The *Challenger* criss-crossed the North Atlantic, swerved down through the South Atlantic, and eastward into the Antarctic Ocean. Leaving the Antarctic, it continued its way to Australia and the western Pacific islands, eastward to the Hawaiian Islands, on through the Straits of Magellan, and finally back to England where it landed on May 24, 1876. In three years the *Challenger* had sailed 68,890 miles and made 362 "oceanographic stations," gathering data on weather conditions, surface currents, water temperature, water composition at various depths, marine organisms, and bottom sediment. Expedition scientists charted oceanic topography on the ship's track, netted and classified 4,717 new species of marine life, and took a depth measurement of 26,850 feet in what came to be known as the *Challenger Deep*.

The official report of the *Challenger* expedition, which filled 29,500 pages in 50 volumes and took 23 years to complete, placed the science of oceanography on a firm footing. The last volume appeared in 1895—three years after Robert E. Peary discovered that Greenland was an island, not part of a polar continent; one year before a Norwegian explorer, Fridtjof Nansen, proved that no such continent existed; 14 years before Peary became the first man to reach the North Pole; 17 years before Roald Amundsen, another Norwegian explorer, reached the South Pole.

Although the *Challenger* expedition is one of the great achievements in the history of scientific exploration, it was nonetheless a candle in a vast darkness. A quarter of a century later, oceanographers had gained a general picture of the earth's oceans; and another quarter of a century after that, oceanography entered upon a new era with the *Meteor* expedition of 1925–1927.

During the half-century from the *Challenger* to the *Meteor,* oceanographic research had consisted mostly of isolated and widely scattered observations. The emphasis was on the amount of territory covered rather than on systematic research. The most notable exceptions to this rule were the *Norwegian* studies of the North Atlantic and the Norwegian Sea begun shortly after the turn of the century. When the German South Atlantic Expedition embarked on the *Meteor* in 1925, it undertook what had never before been attempted—the systematic study of a single ocean. In a period of 25 months, the *Meteor* crossed the South Atlantic more than a dozen times. The data obtained—gathered day and night, in all weather and in all seasons—included some 70,000 soundings of the ocean depths. It was the *Meteor* expedition that first revealed the ruggedness of the ocean floor.

Although oceanographic research became increasingly thorough after the *Meteor* expedition, there was nothing to equal it for another thirty years. It was not until the International Geophysical Year of 1957–58 that oceanographers once again undertook to survey an entire ocean. In 1957 one British and three American ships began crisscrossing the Atlantic from north to south. In the course of their survey, they repeated a number of the actual crossings and many of the stations made by the *Meteor*. One very important result of this joint effort was the first atlas of the Atlantic Ocean.

United States Oceanographic Research

Although Maury is considered "the father of American oceanography," United States support of oceanographic exploration really began in 1877 with the voyages of the research vessel *Blake,* a steamer belonging to the U.S. Coast Survey (now the U.S. Coast and Geodetic Survey). The scientific staff was under the direction of Alexander Agassiz, son of the famous Swiss naturalist Louis Agassiz.

The younger Agassiz, who was responsible for two of the volumes from the *Challenger* expedition, invented a number of sampling and measuring devices, including the "Agassiz trawl," which represented a great advance over earlier trawls. In the last half of the nineteenth century, Agassiz contributed greatly to the growth and development of oceanographic research in this country.

While it would be impractical to attempt a complete listing of the men and ships that have taken part in the United States program of oceanographic research since the first voyage of the *Blake,* it may be of some interest to the reader to know about a few of the United States marine laboratories and Federal agencies currently engaged in marine studies.

Lamont Geological Observatory. The Lamont Geological Observatory of Columbia University was founded at Torrey Cliff, Palisades, New York, in 1949, and is under the direction of Dr. Maurice Ewing. Here marine scientists prepare for voyages aboard the Observatory's chief research vessel *Vema.* The *Vema* is a 202-foot, three-masted schooner with a displacement of 743 tons. Using auxiliary diesel power, she can cruise at a maximum of 8 to 10 knots (nautical miles per hour) and carries a crew of 19 and a scientific staff of 17 (see Fig. 1-1).

Scripps Institution of Oceanography. Scripps Institution of Oceanography of the University of California came into existence on October 13, 1925, at a site overlooking the Pacific Ocean just north of La Jolla, California. Until very recently,

Lamont Geological Observatory

Fig. 1-1. The *Vema,* the main research vessel of the Lamont Geological Observatory.

this laboratory was directed by Dr. Roger Revelle. Scripps is a marine laboratory devoted to the broad spectrum of oceanographic research. The Institute's first long expedition, in 1951, was made by the steel-hulled *Horizon,* a 143-foot, ocean-going tug with a displacement of 900 tons and a maximum speed of 12½ knots. The *Horizon* carries a crew of 19 and a scientific staff of 16 (see Fig. 1-2).

Woods Hole Oceanographic Institution. Chartered in 1930 as a private, non-profit organization devoted to scientific study of the world's oceans, the Woods Hole Oceanographic Institution operates one of the first modern ships designed specifically for deep-sea oceanographic research. This ship, the recently completed *Atlantis II* (see Fig. 1-3), is 210 feet long, has a

Scripps Institution of Oceanography

Fig. 1-2. The *Horizon,* research vessel operated by the Scripps Institution of Oceanography.

cruising speed of 12 knots, and can accommodate a crew of 28 and a scientific party of 25 (the original *Atlantis,* built in 1931, can carry a scientific staff of only nine men). Dr. Paul M. Fye has been director of the Institution since 1958.

United States Navy Oceanographic Office. The primary mission of the United States Navy Oceanographic Office is to improve the combat readiness of the fleet, but basic research performed as part of this assignment is of fundamental importance to all oceanographers. Rear Admiral Denys W. Knoll, Oceanographer of the Navy and Commander of the U.S. Navy Oceanographic Office, directs this office's responsibility for ocean charts, navigation reports, and part of the vast military oceanography program of the U.S. Navy.

United States Coast and Geodetic Survey. The United States Coast and Geodetic Survey, directed by Rear Admiral H. Arnold Karo, was created by an act of Congress on February 10, 1807, after President Jefferson requested a survey of the coasts of the United States. As the primary agency responsible for hydrographic surveying, the Coast and Geodetic Survey operates about 17 ships. The largest of these are the sister ships *Pathfinder* and *Explorer,* and *Pioneer,* a former U.S. seaplane tender. The *Pioneer* is 311½ feet long, has a displacement of 2600 tons, a maximum cruising speed of 18½ knots. It carries a combined crew and scientific staff of 118 men (see Fig. 1-4). Although underwater mapping is their primary responsibility, most of the Coast and Geodetic Survey vessels are equipped to gather a wide range of oceanographic data.

It should be pointed out that the above list is only a sampling of the total United States effort in oceanographic research. Many other federal agencies and educational institutions are investigating specific phases of the geology, chemistry, physics, and biology of the oceans. The United States "fleet" of oceanographic research vessels now numbers 45 ships. Even

Woods Hole Oceanographic Institution

Fig. 1-3. The *Atlantis II,* the newest research vessel operated by the Woods Hole Oceanographic Institution.

Official U.S. Coast and Geodetic Survey Photograph

Fig. 1-4. The *Pioneer,* a hydrographic research vessel operated by the U.S. Coast and Geodetic Survey.

so, this is less than half the number of such vessels that the United States will need in the years to come.

The National Academy of Sciences' Committee on Oceanography, which is responsible for the long-range planning of United States oceanographic research programs, has recommended the construction of 70 new research vessels. The cost would be something over half a billion dollars, or about 2.5 percent of what it will cost us to send a man to the moon. For many reasons—scientific, technological, and military—this seems a small price to pay.

2

The Chemistry of the Sea

Some Properties of Water

We are so accustomed to thinking of sea water as "salt water" that we often forget that the earth's oceans are more than 95 percent pure water. The actual figure is 96.5 percent. For this reason alone, oceanographers would be interested in the properties of water even if it were an "ordinary" substance.

But far from being ordinary, water is one of the most remarkable chemical compounds found in nature. First of all, it is one of only three substances that occur naturally in liquid form on earth. (The other two are oil and free, or uncombined, mercury.) The liquid state of water under normal conditions is explained by the fact that it has a boiling point of 100° C (212° F). This is unusually high considering the type of compound that water represents. The boiling points of related hydrogen compounds (such as ammonia, NH_3, and hydrogen sulfide, H_2S) are lower by at least 80 centigrade degrees.

The high boiling point and other "abnormalities" of water all stem from the particular structure of the water molecule itself (H_2O). The two hydrogen atoms form an angle of about 105° with the oxygen atom (see Fig. 2-1). Consequently, the molecule is *polar*. That is, the electric charges of the atoms do not completely cancel one another out; the oxygen "end" of the molecule is slightly negative with respect to the hydrogen atoms.

Because of this polarity, water molecules cluster together in aggregates of two to eight individual molecules. These aggregates are held together by what are known as *hydrogen bonds.* When the hydrogen atom of one molecule is attracted to the oxygen atom of another, the hydrogen atom forms a "bridge"

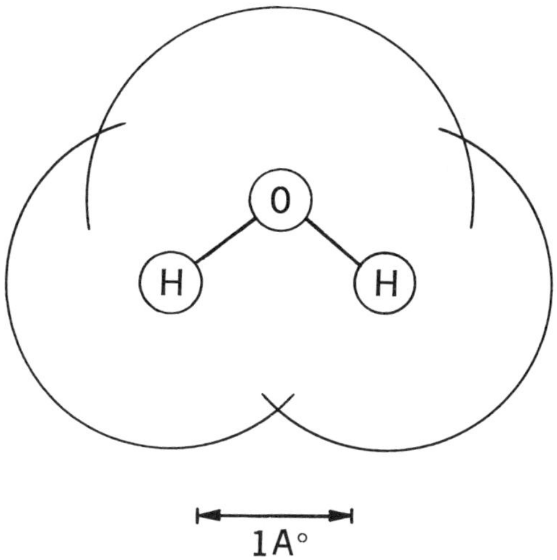

Fig. 2-1. Water molecule. The angle of about 105° between the hydrogen atoms accounts for the polarity of the molecule. If the angle were 180°, the molecule would be nonpolar.

between the two oxygens. This "bridge" is the hydrogen bond. Even at the boiling point, almost 80 percent of the molecules are still bound together by hydrogen bonds (mostly in two-molecule aggregates). At the freezing point, more than 90 percent of the molecules are bound together in aggregates. Thus, in order for water to boil, a large number of hydrogen bonds must first be broken. This requirement results in the higher-than-expected boiling point. If water consisted entirely of single, unbound molecules it would boil at about –80° C (–112° F). That is, the liquid we know as water would be a gas! Needless to say, it is extremely doubtful that any form of life could have gotten started under such conditions.

The peculiar structure of the water molecule also accounts for the familiar observation that ice floats in water. All the water aggregates take up somewhat more space than would the single molecules composing them. Relatively speaking, the bulkiest aggregate is the one containing eight molecules, which has the shape of a triangular prism (the basic crystal structure of ice). The other aggregates are chainlike molecules.

As the temperature decreases, the proportion of bulkier aggregates increases. This increase becomes apparent in the behavior of water at a temperature of 4° C (39° F). At that temperature, the *density,* or mass per unit volume, of water is greatest. Above 4° C, the density varies with temperature in a normal manner: it decreases with rising temperature and increases with falling temperature; this is the same as saying that water expands when heated and contracts when cooled. But below 4° C, water actually expands as it cools. And at 0° C (32° F), the volume of a block of ice is about 10 percent greater than the volume of water (at 4° C) from which it froze. (Most of this expansion occurs at the freezing point.) Therefore, ice floats, because it is less dense than the surrounding water.

Water is one of the very few substances that expand on freezing. It is thus capable of exerting tremendous pressure once it begins to freeze—up to 30,000 pounds per square inch under the right conditions. Such pressures are sufficient to widen cracks in rocks and split them apart. Consequently, water plays an important part in the weathering of rock.

Even more important in the process of weathering and erosion is the dissolving power of water. The ability to dissolve many substances and often in large quantities is also largely due to the polar structure of the water molecules. As a result, many substances, and particularly salts, enter into solution in water—instead of settling out, as they would tend to do in a nonpolar liquid. It is estimated that the earth's rivers wash some 400 million tons of dissolved salts and other materials into the oceans in a single year. These materials, of course, are carried to the rivers by the runoff of rain from the land. It is this process and the great dissolving power of water that have made the oceans "salty."

But rain also contributes to the composition of the oceans by dissolving gases (including carbon dioxide and oxygen) and other substances from the atmosphere. Moreover, the volume of precipitation over the oceans is about three times what it is over the continents, and at least 10 times the volume of water discharged by rivers. (The rivers of the earth discharge about 27,000 cubic kilometers—roughly 6500 cubic miles—of water into the sea in one year.) As one might expect, oceanic climate is an important factor in determining the composition of the oceans.

The Chemical Composition of the Oceans

The addition of salts and other materials to the oceans has probably been a continuous process ever since rain first began to fill the ocean basins. Today the oceans contain about 50 million billion (50 followed by 15 zeros) tons of salt. And the process continues today.

The marine chemist is primarily concerned with the composition and distribution of the ocean waters. Dissolved salts, temperature variations, the kind and amount of dissolved gases, types of mineral suspensions, and the radioactivity of sea water and sediments are just a few of the research subjects in chemical oceanography.

For the chemical oceanographer, the fundamental property of sea water is its *salinity*—the concentration of dissolved salts. He is therefore interested also in the variation of salinity with depth and from one part of the ocean to another. However, these differences, though significant, are quite small. In addition, the ocean currents, both vertical and horizontal, serve to eliminate local variations by distributing the dissolved materials throughout the huge mass of the oceans. Numerous chemical analyses of water samples from different parts of the oceans have shown that the composition of sea water is constant within certain narrow limits.

One thousand grams of sea water contain, on the average, 35 grams of dissolved salts. The oceanographer expresses this by saying that the salinity is *35 parts per thousand* (or 35 ‰). That is, dissolved salts account for 3.5 percent of the weight of sea water. By contrast, average drinking water in the United States usually has a salinity of 0.05 ‰.

Of the many salts present in ordinary sea water, sodium chloride (NaCl), or common table salt, is the most abundant. Table 2-1 gives the relative abundances of the 10 most com-

Table 2-1. Composition of Sea Water.

Element	Parts per thousand
Chlorine (Cl)	18.98
Sodium (Na)	10.56
Magnesium (Mg)	1.27
Sulphur (S)	0.88
Calcium (Ca)	0.40
Potassium (K)	0.38
Bromine (Br)	0.065
Carbon (C)	0.028
Strontium (Sr)	0.013
Boron (B)	0.005

mon elements in sea water. The "salts" referred to in the term *salinity* are combinations of these elements with one another and with other elements. For example, carbon occurs not only in dissolved carbon dioxide (CO_2) but also in the negatively charged bicarbonate ion, HCO_3^-). Actually "salinity" is a slightly misleading term, for it includes *all* dissolved substances (carbonates, for example). The "true salinity," or concentration of salts alone, is somewhat less; but for our purposes the difference is not significant.

The figures in Table 2-1 show that the 10 most common elements account for about 33.5 ‰ of the average salinity. The remaining 1.5 ‰ includes many elements, among them uranium, copper, aluminum, manganese, silver, and gold. Indeed, oceanographers feel fairly certain that all the natural elements are to be found in the sea, though in some cases the concentration must be extremely small. As we shall see, however, the oceans are even now a source of several important elements present in quite low concentration.

Water-sampling Techniques

Benjamin Franklin once placed a Fahrenheit thermometer in a bucket of sea water to determine whether or not he was in the warm current of the Gulf Stream. Unfortunately, acquiring information about water conditions below the surface is much more complicated. In order to obtain measures of salinity and other data, oceanographers must, obviously, obtain samples of sea water at different depths. But how does one get a water sample from well below the surface without contaminating it with other water as the sampler is drawn back to the surface? Then, too, if subsurface currents or the motion of the ship prevent the sampler from descending straight down, how does one determine the depth at which the sample was taken? And, finally, how does one correct for the temperature increase that is caused by the pressure of the water above the sample? One of the most widely used devices for solving these problems is the *Nansen bottle,* named for the Arctic explorer who developed it around the turn of the century.

The Nansen bottle is a hollow cylinder, about 20 inches long, with shut-off valves at both ends. At least two *reversing*

Official U.S. Navy Photograph

Fig. 2-2. A sailor removes a Nansen bottle from the hydrographic wire.

thermometers are attached to the outside of the sampler, and both ends of the bottle are clamped to a hydrographic wire (see Fig. 2-2). As the bottle is lowered, water flows through the cylinder. When the bottle has been lowered to the desired depth, the oceanographer drops a *hydrographic messenger*, or small weight, down the wire. The messenger hits and trips a release on the top clamp. As the bottle overturns in the water (still held by the bottom clamp), the shut-off valves are closed automatically and the thermometers are broken at constrictions in the thermometer tubes.

Breaking the thermometer tube at the base prevents any change in the temperature record while the bottle is drawn to the surface. One of the thermometers is sealed in a thick glass tube and thus is not sensitive to an increase in water pressure. This thermometer records the actual temperature. The other, which is not protected, gives a higher temperature reading because water pressure forces the mercury higher up the tube. Since the density of water is known, the difference between the true and the pressure-induced temperature reading can be used to calculate the depth at which the sample was taken.

Frequently, a number of Nansen bottles are placed at intervals along the hydrographic wire. The reversal of the first bottle releases a hydrographic messenger, which triggers the reversal of the second bottle, and so on down the wire until all the bottles contain sealed samples. Chemical analysis of the samples begins as soon as they are taken on board ship. This is necessary for accurate measurements of, among other data, the oxygen content of the samples. Other measurements are often delayed until the research vessel returns home and the samples can be examined in a fully equipped laboratory.

Because some important chemical elements and gases are present in very low concentration, it is necessary to obtain large volumes of sea water in order to make accurate measurements. The Nansen bottle holds only 1.5 liters, or slightly more than one and a half quarts. One device for obtaining the required amount of water is the "water barrel." This can be used to collect 220-liter samples (about 57 gallons) from any depth desired (Fig. 2-3). The principle of operation is similar to

Lamont Geological Observatory

Fig. 2-3. Technicians prepare to bring a full water-barrel aboard the *Vema*. The water in the barrel will be analyzed for trace elements and dissolved gases.

that of the Nansen bottle. And, as in the case of the smaller samples, processing begins as soon as the water barrel is swung aboard ship.

Salinity, Temperature, and Dissolved Oxygen

Data obtained from water samples provides a wide variety of information about the oceans. It was mentioned earlier, for example, that the salinity varies from one part of the oceans to another. The salt content of surface water in the Arctic Ocean, for instance, is about 33 ⁰/₀₀, whereas the salinity in the Persian Gulf is generally about 40 ‰. Moreover the salinity of surface water in the Atlantic Ocean, and particularly in the Northern Atlantic, is significantly higher than it is in the Pacific Ocean. A number of factors are responsible for this difference, including atmospheric processes and salt transport by ocean currents.

Generally speaking, where evaporation of water from the ocean surface is high, as in tropical regions, the salinity is higher than average. In the polar regions, where evaporation is low, the salinity is also lower than average. In the open ocean, unusually low salinity may be caused by the presence of fresh water from the discharge of large rivers and from melting glaciers reaching the shoreline.

To observe one aspect of the relationship between salinity and weather, one has only to go to the beach on a day when a good wind is blowing off the water. After a short time, you can taste the tiny salt crystals that have collected on your lips. The spray thrown up by waves breaking on the shore (and further out as well) has evaporated, liberating minute salt particles into the atmosphere. Meteorologists now think that these salt particles may perform a natural cloud-seeding function. For in areas where the concentration of salt in the atmosphere is higher than average, there is a correspondingly higher rainfall.

Measurements of salinity, combined with data on temperature and oxygen content of water samples, help marine chemists and physical oceanographers to identify the source of large water masses, and thus to chart their movement. Salinity and temperature both affect the density of sea water, as does pressure. The density increases with increasing salinity, with decreasing temperature, and with increasing pressure. If water at the surface becomes denser through evaporation (i.e., increased salinity) or through cooling, it will sink. Its density also will increase with depth; that is, with increasing pressure. Such density variations play a very important part in the circulation of ocean water.

In passing, it should be noted that the increase of density with pressure, though significant, is quite small. Even in the deepest part of the ocean, the density is no more than a few percent greater than the density of surface water—the compressibility of water is very slight. This means that a wrecked ship sinks all the way to the bottom, contrary to the rather common myth that it cannot do so. It would remain suspended somewhere below the surface only if the water density at some point equaled the density of the iron or steel of the ship. (The density of iron, for example, is about seven times that of water.) Since water is almost incompressible, this never happens. No matter how light it may be, almost any object heavy enough to sink below the surface will eventually reach the ocean floor, though it may take a long time to do so.

The amount of dissolved oxygen in a given water mass depends in part on the initial concentration and on how much oxygen is used up by marine animals and plants. A liter of sea water at the surface may contain a maximum of 10 milliliters of dissolved oxygen. As the water sinks, the oxygen is gradually used up by marine organisms. Consequently, the oxygen content of water in the deep ocean is an index of the water's "age." During the International Geophysical Year, for example, oceanographers found that the oxygen content of "bottom water" in the Antarctic and of "deep water" in the North Atlantic had decreased since the previous measurements, made 30 years earlier. The salinity and temperature, on the other hand, had remained unchanged. This was very strong, but not conclusive, evidence that the water of the huge masses had not been replenished at that time. One might say that salinity, temperature, and dissolved oxygen are the "fingerprints" of water masses.

Idealized average conditions of temperature, dissolved oxygen, and salinity in a large body of water might look like those shown in Figure 2-4. The top of each graph represents the ocean surface, or zero depth. The relative values of each factor decrease to a minimum at the left side of each graph.

The temperature-depth graph shows that temperature is uniform throughout the surface layer, but decreases rapidly below the base of the surface layer. This layer of rapid temperature change over a small range of depth is called the

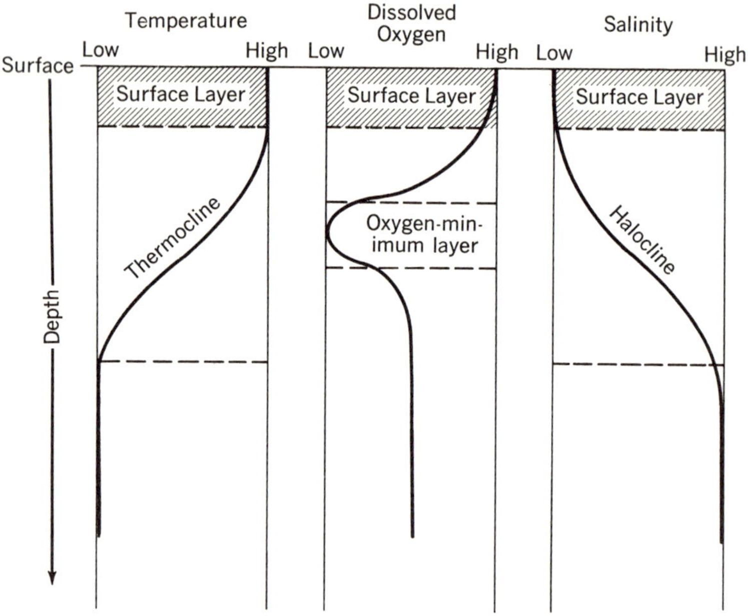

Fig. 2-4. Idealized graphs of changes of temperature, dissolved oxygen, and salinity with increasing depth in the open ocean.

thermocline. Temperature is fairly constant for all depths below the base of this layer.

In the second graph, the amount of oxygen dissolved in sea water is shown to decrease from the surface downward to a layer at which little, if any, oxygen is present. This is the *oxygen-minimum layer,* and it is usually found within the thermocline. The amount of dissolved oxygen increases rapidly below the oxygen-minimum layer and then becomes fairly constant with increasing depth.

As the third graph shows, salinity is uniformly low in the surface layer, below which is found the *halocline.* The rapid increase of salinity with depth in the halocline is followed by fairly uniform salinity at all depths below. For convenience, the thermocline and halocline layers have been shown within the same depth range. In the ocean, however, these layers may extend over different ranges of depth.

To generalize, the major thermocline, halocline, and oxygen-minimum layers are usually found within 500 meters (1640 feet) of the surface. These are more or less permanent conditions found in the deep oceans, although local, daily, and seasonal influences, especially at the surface, frequently cause variations in sea-water chemistry. When you go to the beach and dive into the water, you may feel a sharp decrease in temperature just a few feet below the surface. This is a shallow thermocline, caused by wave action and the heating of surface water by the sun. At night you would probably search in vain for such a shallow thermocline. Similarly, rainfall on the ocean surface creates a shallow minor halocline; and local wave action causes a temporary increase in the amount of dissolved oxygen in the surface layer.

The Importance of Dissolved Chemicals in Sea Water

It has already been mentioned that the oceans are a source of many substances that are valuable to man. Some of these substances are important to man in their own right, while others are important for the part they play in the life of the sea.

Among the latter, for example, are the mineral nutrients required by the marine plants that are the supporting base of the "food-chain" in the oceans (see Chapter 7). Some invertebrate marine animals extract dissolved silica from ocean water. Still others use calcium carbonate to build their shells. These in turn have provided the "building blocks" of many tropical islands and coral reefs.

The importance of fish in the food-chain is largely dependent on the many dissolved chemicals they utilize. Iodine, for instance, is absorbed by fish and eventually finds its way into the human diet. This mineral is absolutely essential to the proper functioning of the thyroid gland. Fish also incorporate phosphorus through the food-chain.

Phosphorus is an essential "mineral" in the human diet. For this reason, most of the phosphate mined in the United States is used in fertilizers. Unfortunately, a great deal of phosphorus is leached out of the soil by rain and carried away by rivers. Of the more than three million tons washed into the sea every year, only about two percent is returned by natural processes. Eventually, some means will have to be devised for mining phosphorus from the oceans. Sediments containing significant quantities of phosphorus have recently been discovered off both coasts of the United States as well as in many other coastal areas all over the world.

Two important chemicals at present being mined from the oceans are magnesium and bromine. All of the magnesium produced in the United States, and more than three-fourths of the world's bromine is extracted from sea water. Magnesium is a light metal that has a number of industrial uses. Bromine is a *halogen* ("salt-former") used chiefly in dyes, photographic chemicals, and as an additive in gasoline.

Both magnesium and bromine are relatively "abundant," as Table 2-1 shows. Less abundant are such useful elements as uranium, copper, manganese, aluminum, silver, and gold. Like phosphorus, both manganese and aluminum may eventually be mined from sediment deposits (see Chapter 6). It has been estimated that the oceans contain about 20 billion tons of uranium, and 15 million tons each of copper and manganese. Silver and gold are much less abundant. According to one estimate, one cubic mile of sea water contains 23 pounds of dissolved gold. At the United States market price of 35 dollars an ounce, a cubic mile of sea water is worth 12,880 dollars for its gold alone. On the other hand, it would cost a great deal more than that to mine the gold.

In the future, it may be possible to extract a number of chemical elements from the oceans on an economic basis. If so, such an undertaking will probably be coordinated with the desalination of sea water for human and agricultural needs. Man has, of course, been getting most of his table salt from the oceans for thousands of years. But this process is much simpler than that of getting fresh water from salt water. The present cost of production is about one dollar per gallon. The need for cheaper production is becoming increasingly urgent.

According to current estimates, the demand for fresh water in the United States will exceed the supply by 1980.

The presence of dissolved chemicals in sea water also raises fundamental questions about the origin of the oceans and about chemical and physical processes that take place in them. These questions are the particular interest of the geochemist. Geochemistry has developed as a specialized field because many questions about the earth's chemical makeup are concerned with geological processes.

For example, there is the question of the age of the oceans. Most oceanographers believe that the oceans began to fill quite early in the earth's history—as soon as it had cooled sufficiently for the water vapor in the atmosphere to condense and fall as rain. On this theory, the oceans filled gradually, reaching their present level about 500 million years ago. Since that time, the total volume has increased only slightly. And almost all of the salt in the sea would have come from runoff from the continents.

But if the ocean salt originally came from the continents, why aren't the oceans saltier than they are? To put the question another way, are the oceans (or ocean basins) as old as the continents? For the geochemist, the problem is to determine the source of the water that now fills the ocean basins. Did it all come from the atmosphere? Or did part of it, as has been suggested, come from volcanic eruptions?

Similarly, there is a need for more detailed understanding of the processes by which chemical substances are distributed in the ocean. Some substances like sodium show a strong tendency to remain in solution, while others, aluminum for example, form compounds that are deposited in sediments. Potassium, on the other hand, is a soluble substance that does not remain in solution. More than 97 percent of the potassium washed into the sea is eventually deposited in sediments. Conversely, the weathering of continental rock does not account for all the chlorine, bromine, sulfur, and boron found in the ocean. Apparently, a large part of these substances must have entered the ocean by way of the atmosphere and from volcanic eruptions.

The selective deposition of certain chemical elements is extremely important in the life of the sea. For instance, copper, lead, molybdenum, and mercury are highly poisonous to living organisms. Fortunately, the concentrations of these and other poisonous elements are very low. If they were not selectively deposited in sediments, however, their concentrations in sea water would be thousands of times greater. The same holds true of some radioactive elements, such as radium. The concentration of radium in sea water is about 10,000 times less than it is in continental rock; but the radium content of ocean sediment is up to 20 times greater than that of continental rock. Whether this selective deposition would have the same effect on radioactive wastes from nuclear reactors is an important question that remains to be answered.

Oceanographers, geochemists, and meteorologists are also interested in the amount of carbon dioxide stored in the oceans, and in variations of the carbon-dioxide content. The chief reason for their interest is the possible effect of carbon dioxide on the climate. With the advent of the Industrial Revolution, man began to pour huge amounts of exhaust and smoke into the atmosphere, thereby adding a great deal of carbon dioxide to the air. Apparently, most of this has been absorbed by the oceans.

The earth's surface constantly reradiates heat derived from sunlight and other solar energy. Part of this heat is absorbed by carbon dioxide in the atmosphere. An increase in the amount of carbon dioxide would, presumably, increase the amount of heat absorbed by the atmosphere. Considering the amount of carbon dioxide that industry is expected to add to the atmosphere in the next 100 years, the earth's temperature might well increase by several degrees—even though a large part of the carbon dioxide would be dissolved in the oceans. Such an increase in temperature would have far-reaching effects on the climate, the polar ice caps, and oceanic circulation. But even a smaller temperature rise might lower the oceans' capacity for absorbing carbon dioxide, thereby increasing the amount of atmospheric carbon dioxide—this could be the beginning of a vicious cycle. (The temperature rise would be self-limiting, however: an increase in the amount of atmospheric water vapor would produce more clouds, which in turn would reflect more of the sun's radiation back into space and thus cool the earth.) It turns out, moreover, that temperatures have been increasing since the middle of the last century. It is not known, however, to what extent, if any, atmospheric carbon dioxide has contributed to this temperature rise.

3

Oceanic Circulation

Surface Currents of the Oceans

Ancient peoples considered the sun to be a god—the giver of light and warmth. We now realize that the sun is not a god, but energy from the sun is nevertheless of fundamental importance to life on earth.

As the source of more than 99 percent of the energy reaching our planet, the sun powers the circulation of air in the atmosphere. Wind patterns, set up in the lower layer of the atmosphere (the *troposphere*) by energy drawn from the sun and from the earth's rotation, push the surface waters of the ocean into huge "circulation cells." These cells of slowly rotating surface and near-surface water are called *gyres*—mighty, circular rivers of water flowing in the oceans. One huge gyre and many smaller ones are found in each ocean area; they are present throughout the year, although they show slight seasonal shifts of position.

Driven by the winds and pulled by friction between water particles, the direction of rotation varies. In the North Atlantic and North Pacific Oceans, the major gyres rotate in a clockwise direction; in the South Atlantic and South Pacific, rotation is counterclockwise (see Fig. 3-1). This difference in direction of rotation results from the rotation of the earth on its axis. To illustrate, a particle moving north from the equator experiences a force that deflects it to the right; that is, the particle tends to move north *and* east. A particle moving south from the equator is deflected to the left; it travels south *and* east. This deflective force, or *Coriolis force,* is zero at the equator and increases toward the poles. The Coriolis force largely determines not only the direction of gyres, but also the direction of the prevailing winds in the atmosphere.

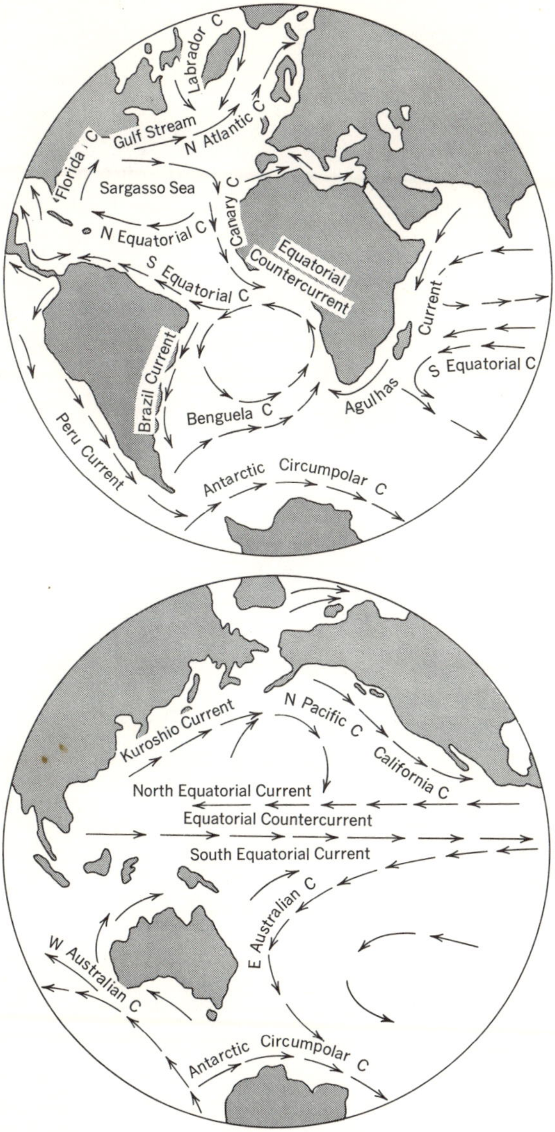

Fig. 3-1. Major gyres and currents of the world are represented on this map (see also Fig. 3-2). Note that the major gyres rotate in a clockwise direction in the northern hemisphere, and in a counterclockwise direction in the southern hemisphere.

Along the equator, the currents of both major gyres flow west, piling water up against the eastern coast of the South American continent (see Fig. 3-1). Because the prevailing trade winds in the equatorial region are weak, this piled-up water in effect flows back downhill, or eastward. This *counter* flow of water, which separates the two major gyres from each other, is known as the *Atlantic Equatorial Countercurrent.* In the Pacific equatorial region also, the predominant flow is westward and the *Pacific Equatorial Countercurrent* runs eastward.

Portions of each of the major gyres have been given the names of the islands, countries, and geographic features near which they flow. Thus, the predominant current just north of the equator is called the *North Equatorial Current,* and in the North Atlantic Ocean we find the *Florida Current,* the *Labrador Current,* and the *Canary Current.* Each of the oceanic gyres affects the climate of the land masses that bound it. The North Atlantic Ocean gyre is one of the best mapped surface-circulation systems.

North Atlantic Ocean Circulation

One of the main currents in the North Atlantic Ocean gyre is the famous *Gulf Stream.* For the sake of convenience, we can call the North Atlantic's entire surface circulation the *Gulf-Stream gyre.*

On September 13, 1492, Christopher Columbus entered in his logbook the first written account of a strong current in the western North Atlantic Ocean. Later, in 1513, Ponce de Leon's three ships nearly lost a bout with the Florida Current. Heading south toward Tortugas from a point near what is now called Cape Kennedy, and with a good breeze filling their sails, the crew noticed that the ships were actually traveling backward! Learning from this and similar incidents, Spanish explorers and merchantmen found that they could return to Europe faster if they sailed north from Cuba, along the Florida Coast, and northeast toward France. Then, more than halfway across the Atlantic, they would drop southward to follow the Canary Current home to Spain.

In spite of this accumulated practical experience, it remained for Benjamin Franklin to study and name the Gulf Stream. As Deputy Postmaster General of the British Colonies in North America until 1774, Franklin had been asked in 1769 why the four packet ships carrying mail from Falmouth, England, to New York required seven weeks to make the journey across the North Atlantic, while American merchant ships returning heavily laden from London to Rhode Island were able to make the trip in five weeks.

Thinking that a mistake had been made, Franklin sought out a Nantucket sea captain, who reported a swiftly flowing stream of water in the ocean that had to be avoided on the passage from England to America but that would speed the voyage from America to England. The captain drew a chart of the stream and reported that it was well known to American whaling captains because whales kept near the sides of the stream of warm water but were not found within the stream.

Franklin made several trips between America and England, measuring the water temperature and speed of the current, and noting the presence of what he called "gulf weed." He correctly assumed that the current was at least partially caused by the trade wind blowing across the southern North Atlantic. It was his erroneous belief that all the warm water in the stream came through the Gulf of Florida (now known as the Florida Straits), that caused him to name this current the Gulf Stream.

The Sargasso Sea. In the center of the Gulf Stream gyre, and almost 5.2 million square kilometers (2 million square miles) in area, lies the Sargasso Sea. It was named by Portuguese sailors for the gold-colored seaweed, *Sargassum,* floating in great abundance at and near the surface. This is the "gulf weed" observed by Franklin. Believing the seaweed would hold ships fast, sailors long feared the Sargasso Sea and called it the "graveyard of ships." As the Sargasso Sea is in a belt of latitude known for its lack of strong winds (the *Horse Latitudes*), it is not difficult to imagine how such a legend began. Actually, the sargassum is not thick enough to prevent passage of even the smallest ship.

Marine biologists have found that the blue water of the Sargasso Sea hides a biological desert. Seaweed is just about the only form of life down to a depth of 915 meters (3000 feet), near the center of the gyre.

Currents and Climate. Water warmed in the western part of the North Atlantic carries its heat to the shores of England, Ireland, France, Denmark, and Norway. There, the average winter temperature is 2.8° to 5.6° C (5° to 15° F) higher than it would be without the warmth that is derived from such branches of the Gulf Stream as the North Atlantic Current and the Norwegian Coastal Current. When air warmed by Gulf Stream waters comes in contact with the normally colder air of these countries, a marine hazard—fog arises. The pea-soup fogs of London are well known, and they are common off the Grand Banks of Newfoundland and in other coastal regions near the Gulf Stream.

Another example of the Gulf Stream gyre's effect on climate may be of interest to students of World War II. During the war, it became necessary for the Allies to supply Russian armies with huge quantities of food and weapons. In those days of short-range airplanes, no airlift of supplies was possible: shipments had to be made to some port in Russia. Vladivostok, on the Sea of Japan, was opposite the Japanese mainland and too far from the eastern front. The Black Sea ports of Odessa and Sevastopol were accessible by way of the Mediterranean, but the route was blocked by ships and submarines of the Axis powers; an alternate route through the British-held Suez Canal was lengthy and required rounding the Cape of Good Hope. Leningrad, on the German-controlled Baltic Sea, was icebound in winter. But one port remained. Murmansk, east of Norway on the Barents Sea, is washed by the warm waters of the North Cape Current. If it were not for the Gulf Stream gyre, this port would be completely icebound in winter. A high percentage of the merchant ships traveling from the United States to Murmansk were sunk by the Axis-submarine "wolf packs" lying in wait along this vital supply line. Needless to say, the story of the dangerous "Murmansk Run" has been the subject of many war novels and films.

Current Velocity. Continuing the work begun by Matthew Maury, the United States Navy Oceanographic Office today issues a monthly "Pilot Chart" of the oceans. These charts show average wind velocity and direction, water temperature, surface-current velocities, and much other information of practical interest to mariners. For example, the pilot chart for June shows that the general clockwise rotation of the Gulf Stream gyre has a velocity of about 0.21 to 0.26 meter per second (0.4 to 0.5 knot), and in January it increases to about 0.26 tc 0.31 meter per second (0.5 to 0.6 knot). Around 1800, the German explorer and scientist Alexander Von Humboldt estimated that it would take water in this gyre two years and ten months to make a complete circuit. Measures of current velocities since Humboldt's time have confirmed his original estimate.

Significant departures from the general velocity of current flow are found in the western North Atlantic Ocean, where the North and South Equatorial Currents along the coast of Brazil and the Guianas reach velocities of 0.62 meter per second (1.2 knots) in June, and 1.0 meter per second (2.0 knots) in January. In the Caribbean Sea, the Gulf of Mexico, and along the east coast of the United States during both winter and summer, current flow is at about 0.52 meter per second (1 knot). The swiftest current is found along the east coast of Florida, where the Florida Current reaches a velocity of 1.4 meters per second (2.5 knots) in January.

Measuring Surface Currents. In addition to the records in ships' logs of currents encountered on each journey, there are many devices available for measuring the flow of streams, whether these streams are continental rivers or in the ocean. Colored dyes can be traced for long distances. Current meters that work on the principle of the anemometers used by weather men are also employed to determine current characteristics at specific locations. One of the least expensive ways of measuring average current conditions is by using corked bottles (drift bottles). Released at sea, the bottles will follow the surface currents and eventually drift back toward land. When they are discovered on the beach, the finder fills out the enclosed card and sends it to the oceanographic agency, describing the location and date found.

Surface-Current Studies. Within recent years, anthropologists have intensified their efforts to determine if the route of man's migration to and from the New World might have been by crude boats that simply drifted with the surface currents. There are many popular and scientific reports about the Kon-Tiki expedition, a modern-day attempt to see if such a trip between South America and Polynesia was possible.

Similarly, zoologists and botanists are investigating the possibility of accidental migration, between continents, of animals and plants caught up in a gyre. This is a fundamental question, because the alternative is to assume that land bridges once existed between continents and islands, permitting plants and animals to migrate between continents over dry land. As we do not find such dry-land bridges today, it must be assumed that they sank into the ocean or that the ocean level rose above them. There is sound geologic reasoning to support the latter hypothesis, but much study remains before the exact routes of early migration can be specified.

Deep Circulation of the Oceans

The pattern of surface-water circulation in the oceans is fairly well known and is more easily mapped than is the pattern of deep circulation. As more subtle instrumentation and techniques become available, it may be possible to map the deep circulation in the required detail. The earth's weather and climate influence, and are influenced by, the surface-water circulation. But, in some areas, cold water from below the surface layer is forced upward to form currents of a type called *upwellings.* The exact causes for these upwellings are varied. In some cases, the upwelling is more closely related to deep-circulation mechanisms in the ocean. In the United States, upwellings are common along the northern coast of Florida and along the California coast line. Where such large volumes of cold water are present in the coastal area, they tend to cool the land. The water that is forced to the surface is also usually rich in nutrients that are important to the tiny marine animals and plants known as *plankton* (see Chapter 7). Fish might be expected to concentrate in and near these areas of rapid plankton growth.

Why should there be a movement of ocean water at depths below the surface layer? Certainly there is no wind to push the deep water as is the case with surface waters. But even though winds do not directly affect deep ocean waters, it has been possible to demonstrate in the laboratory that surface winds on a rotating earth will cause a circulation of deep water. Is the circulation pattern of the model greatly different from that to be found on the actual earth? To answer this question, physical oceanographers must chart the average conditions of deep-water circulation in the great ocean basins, an undertaking that will require many years of effort.

The Water Budget. Some of the reasons for assuming the existence of deep-water circulation are a little more obvious than the hydrodynamic-model analysis mentioned above. When we look at the surface water circulation of the North Atlantic Ocean, we see that water is carried into the North Atlantic from the South Atlantic Ocean by the South Equatorial Current (see Fig. 3-2). As the South Equatorial Current approaches the coast of Brazil, it splits into a northerly branch that transports surface water into the North Atlantic. Somehow this northward flowing water must be replaced by an equivalent amount of water flowing back into the South Atlantic Ocean. Since no return flow of sufficient volume has been found in the surface water layer, it must be assumed that the water budget between oceans is balanced by a return flow of some subsurface layer.

Thermal Circulation. There is another reason for assuming a deep circulation of ocean waters. As we saw in Chapter 2, if fresh water at 4° C is warmed, it expands and becomes less dense; if cooled, the water also expands and becomes less dense. Salt water behaves in a slightly different manner. Ocean water of normal salinity (35 ‰) freezes at −1.9° C (28.6° F). This water would be most dense at –3.5° C (25.7° F). Therefore, prior to freezing, the colder this saline water be-

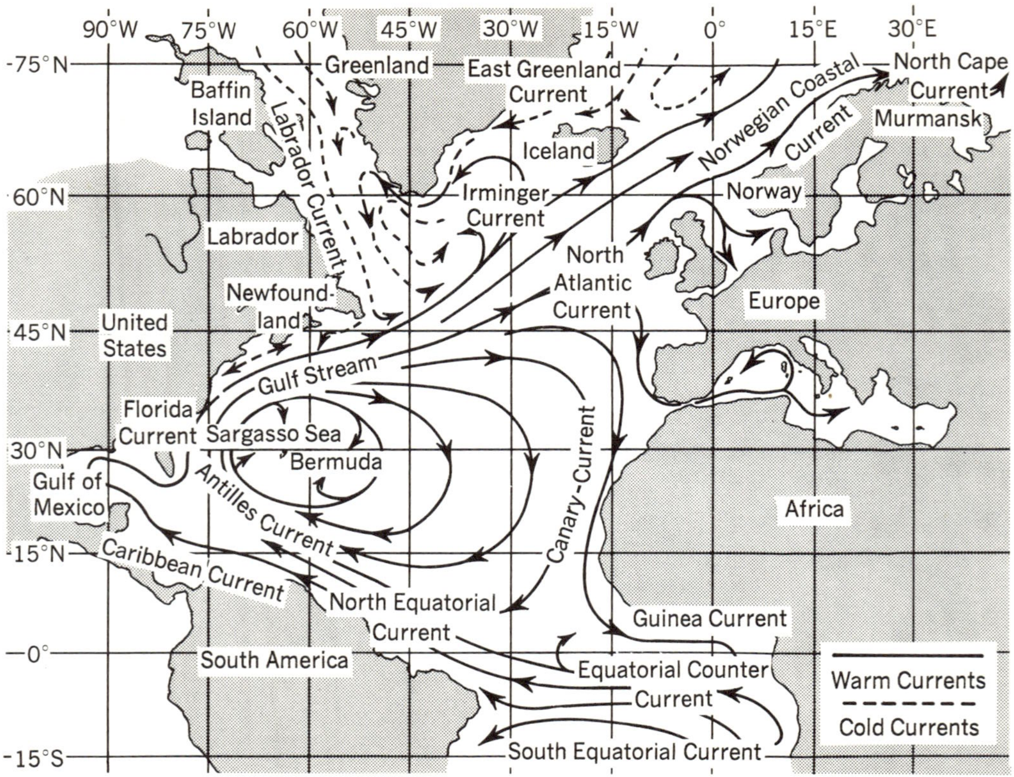

Fig. 3-2. Map of surface currents in the North Atlantic Ocean. Arrows indicate the direction of current flow. The major circulation cell shown on the chart, the Gulf Stream Gyre, turns slowly in a clockwise direction.

comes, the denser it will be and the stronger will be the tendency for it to sink below the surface layer.

Cold ocean water flowing into tropical regions will warm, expand, and become less dense, or lighter. It can be pictured as spreading on the surface of the ocean from the tropics toward the polar regions. As the water travels toward the poles, it is cooled, and this cooling causes the water to contract and become more dense. If the surface water becomes denser than water lying below the surface, it will sink to a depth at which it can spread toward tropical regions. This simple temperature-controlled, convection-cell circulation is called the *thermal circulation* of the oceans.

Haline Circulation. A third kind of evidence of deep-water circulation is related to variations in ocean-water salinity. Since tropical regions are quite warm, there is a great deal of water evaporation from the surface of tropical oceans. As evaporation proceeds, there is an increase in the salt content of the water left behind. These waters are said to have a higher salinity concentration. The salinity increase causes an increase in density of the water and the possibility that the surface water will sink and be replaced by the less dense water lying below. This deeper water is less dense because the salinity concentration is lower.

We have already seen how the temperature-density relationship can cause water to sink in polar regions. But what happens when ocean water freezes? As the temperature drops to the freezing point of ocean water (–1.9° C, or 28.6° F), ice crystals of fairly pure water begin to form. As ice formation continues, the salinity concentration of the water that does not freeze greatly increases and, therefore, its density greatly increases. This dense surface water will sink and be replaced by fresher water, a process that continues until a layer of ice covers the water surface. The salinity-density convection-cell circulation described above is called the *haline circulation.*

Deep Water and Bottom Water. Which of the deep-water circulation mechanisms is dominant in a given water mass? Is it the thermal circulation or the haline circulation, or perhaps both? There is as yet no precise answer to this question. More than one convection cell in each ocean is created by both thermal and haline circulations. Some of the complexity of the problem can be gauged by considering the water masses that would generally be present in an average north-south cross-section through the North Atlantic Ocean. There is a suggestion that a water mass found at a depth of 1500 to 4000 meters (4900 to 13,100 feet) in the North Atlantic Ocean, and traced as far south as 40° south latitude in the South Atlantic Ocean, is formed by the sinking of water in the area just south of Greenland. This water mass is known as the *North Atlantic deep water.* While its salinity of 34.9 ⁰/₀₀ is about the average salinity of ocean water, its temperature–2° to 3° C (35.5° to 37.5° F)–is rather low, which may explain the sinking and southward flow of the waters into the South Atlantic Ocean.

North Atlantic deep water is not the deepest water mass. Below it, over a large part of the North Atlantic Ocean basin, lies a water mass of above-average salinity (36.7 ⁰/₀₀) that has a temperature of 0° C (32° F). This water has been traced from the Weddell Sea area of Antarctica all the way north to about 34° north latitude. This water mass is called *Antarctic bottom water.* Both its temperature and salinity cause sinking and northward flow. In most cases, both temperature and salinity are important factors in the formation of water masses, and the physical oceanographer describes the deep ocean circulation as primarily a *thermohaline circulation.*

Deep Circulation of the North Atlantic Ocean. Model studies and tracing of water masses have resulted in what is very likely a fairly accurate description of the route taken by North Atlantic deep water. The diagram of generalized conditions shown in Figure 3-3 suggests that a thermohaline circulation is initiated when surface water sinks in the area just south of Greenland. The *deep water* that is formed flows southward in a swift stream beneath the Gulf Stream. Part of the deep water flows into the South Atlantic, and part is returned to the north in a deep gyre that rotates in a counterclockwise direction under the clockwise-rotating Gulf Stream gyre on the surface. It has recently been estimated that the volume of flow of *bottom water* from the South Atlantic Ocean

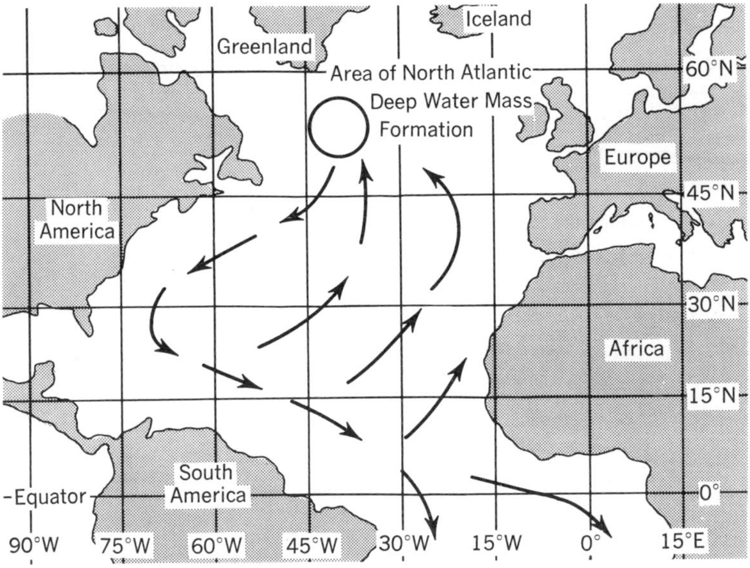

Fig. 3-3. Hypothetical model of the circulation of the water mass known as North Atlantic Deep Water. (*After H. Stommel, 1958*)

is 3 million cubic meters (106 million cubic feet) per second. *Surface* and *intermediate* water from the South Atlantic Ocean accounts for an additional 23 million cubic meters (812 million cubic feet) per second entering the North Atlantic Ocean. It has been suggested that the North Atlantic deep-water mass transports a water volume of 26 million cubic meters (918 million cubic feet) per second into the South Atlantic Ocean, thereby balancing the flow of water between oceans.

Deep Countercurrents. Analysis of current measurements from a few stations in the western North Atlantic Ocean indicate that a deep current does exist. It flows in a direction opposite to the Florida Current portion of the Gulf Stream gyre and for this reason it is called a *countercurrent.* A graph of the relation between current velocity and depth near Cape Romain, South Carolina, is given in Figure 3-4. The graph shows that the Florida Current decreases in velocity from 1.5

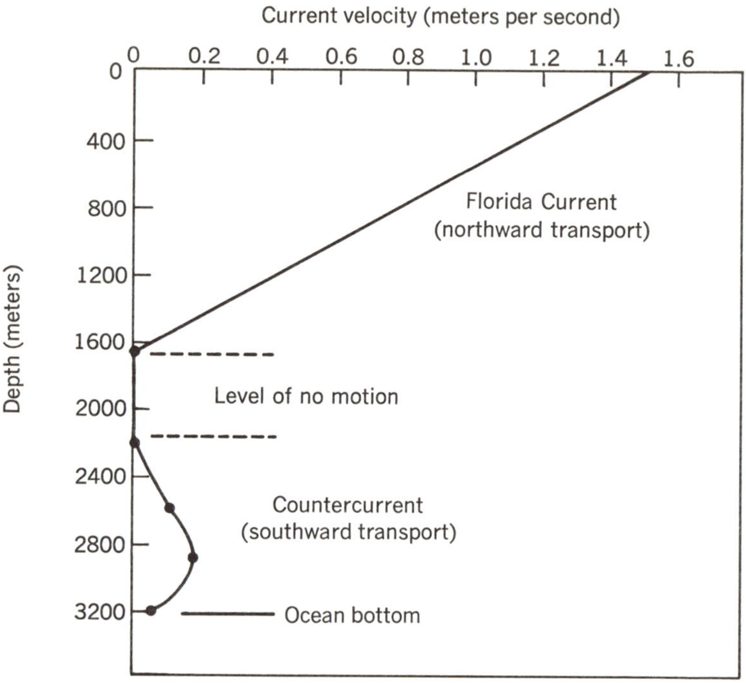

Fig. 3-4. Graph of velocity versus depth in the Florida Current and Countercurrent as measured near Cape Romain, South Carolina (latitude 33°N, longitude 76°W).

meters per second (3.0 knots) at the surface to zero velocity at 1640 meters (5380 feet). There is no measurable horizontal motion down to about 2150 meters (7050 feet). Below this, a countercurrent begins and is found all the way to the ocean bottom at 3200 meters (10,500 feet). This current has a maximum velocity of about 0.18 meter per second (0.35 knot). Maximum velocity is found at about 2840 meters (9315 feet). The amount of water flowing northward in the Florida Current has been estimated at a maximum volume of 70 million cubic meters (roughly 2½ billion cubic feet) per second. In order to put this measurement of Florida Current volume-transport in perspective, it should be noted that the Mississippi River discharges an average of 17,300 cubic meters (611,000 cubic feet) every second into the Gulf of Mexico. In terms of water volume, therefore, the Florida Current is approximately 4000 times larger than the Mississippi River system. The southerly flow of water in the countercurrent below the Florida Current accounts for a volume of about 5 million cubic meters (176.5 million cubic feet) per second, which makes the countercurrent flow about equal to 290 Mississippi Rivers.

Sediment on the ocean bottom is often found to have a rippled surface. Almost 80 percent of a series of bottom photographs taken in the Florida Straits showed asymmetrical ripple marks, which would indicate a flow of water to the south underneath the northerly flowing Florida Current. Rippled sands have also been found at many other locations in the deep oceans. Depths at which rippled bottom has been found are beyond the depths usually associated with ripple formation caused by the rotating motion of water particles in surface waves. Also, their asymmetrical shape is different from the symmetrical ripples usually associated with such motion.

Dense bottom flows of water and sediment (*turbidity currents*), which are not related to the pattern of deep circulation, are also pictured as the cause of rippled bottom seen in photographs. Because some of the ripples have been found on the flat tops of seamounts, areas that turbidity currents cannot reach, it must be assumed that both watercurrents and turbidity currents can produce ripples in bottom sediment.

Undercurrents. Where separate currents flow at various depths, one current may flow parallel to the surface current but in the opposite direction. There may be more currents in this cross-section, each flowing in a different direction. This situation is much more difficult to picture than is the relatively simple current-countercurrent relationship observed in the western North Atlantic Ocean. Oceanographers prefer to term each of the individual currents found beneath the surface an *undercurrent*.

One of the most famous undercurrents, the Cromwell Current, is a Pacific Ocean equatorial undercurrent discovered in 1951 by the late Townsend Cromwell, an oceanographer with the U.S. Fish and Wildlife Service.

The Cromwell Current is a shallow, swift current flowing

eastward along the equator and directly below the South Equatorial Current. As much as 250 miles wide in places, the current is a broad, thin ribbon of flow extending from about 30 meters (100 feet) to 230 meters (750 feet) below the surface. With a maximum velocity of about 1.6 meters per second (3.1 knots), the Cromwell Current is the fastest known current in the equatorial Pacific. The eastward volume of flow is estimated to be 40 million cubic meters (1.41 billion cubic feet) per second, or about one-half the volume flow of the Gulf Stream. This makes it the second largest current in the Pacific, after the Kuroshio, or Japan, Current.

Although the Cromwell Current has been traced from the equator at about 150° west longitude to the Galápagos Islands, 600 miles off the coast of Ecuador, neither end of the undercurrent has been located. However, it probably runs at least 13,000 kilometers (8,000 miles) across the Pacific Ocean—from New Guinea to the coast of Ecuador.

Nor is it known how the Cromwell Current fits into the overall water budget in the Pacific Ocean. During the International Geophysical Year, oceanographers found that the Pacific Equatorial Countercurrent carries more than half again as much water eastward as had been previously thought. The volume of the Equatorial countercurrent, added to that of the Cromwell Current, is three times the theoretical estimates of equatorial flow eastward. But even less is known about the westward flow that must compensate for these two currents. At a depth of about 500 meters (1640 feet), another undercurrent is found flowing to the west beneath the Cromwell Current. But the extent and volume of this current remain to be determined.

Much of the effort of physical oceanographers is now directed to finding and tracing the undercurrents in all oceans. An undercurrent similar to the Cromwell is known to flow beneath the South Equatorial Current in the Atlantic Ocean. Like the Cromwell Current, this current raises questions about the water budget. Oceanographers suspect that there is a corresponding undercurrent in the equatorial region of the Indian Ocean.

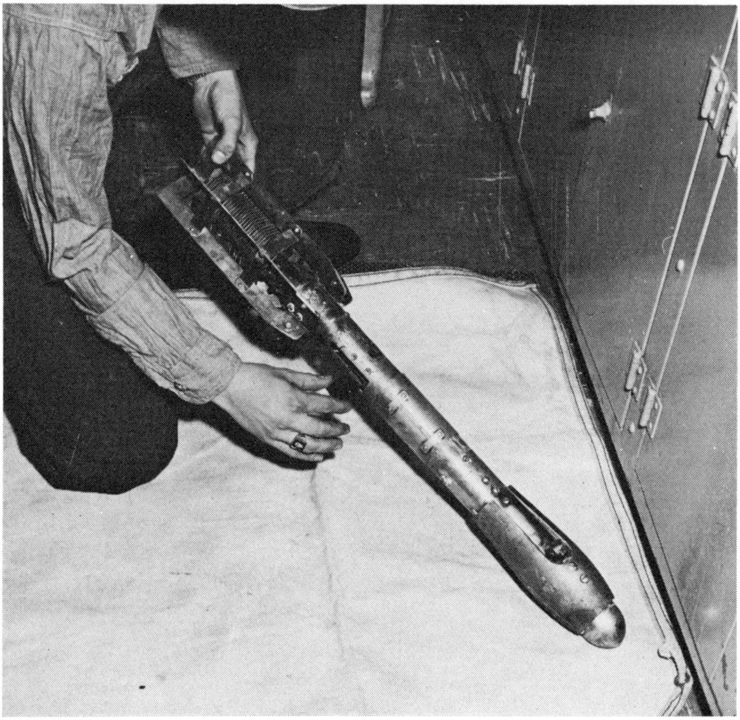

Official U.S. Navy Photograph

Fig. 3-5. A technician inserts a smoked-glass slide into a bathythermograph. Changes in pressure and temperature found as the instrument is lowered beneath the surface are scratched on the slide in the form of a graph.

Measuring the Characteristics of Deep Water. The Nansen bottle, which has already been described in Chapter 2, is used to sample deep water in the oceans as well as water closer to the surface. Where it is desired to obtain only the temperature of ocean water between the surface and a maximum depth of approximately 275 meters (900 feet), a device called a *bathythermograph* is used (see Fig. 3-5). A stylus within the body of the bathythermograph "fish" is activated by a temperature-sensitive liquid in a capillary tube wound around the tail section of the instrument. A "smoked" (carbon-coated) glass slide

Official U.S. Coast and Geodetic Survey Photograph

Fig. 3-6. Roberts type current meter being lowered over the side of the *Pioneer*. This meter can measure current velocities as low as 1/20 of a knot. Current direction with respect to magnetic north is also signaled to the shipboard recorder.

is placed inside the "fish" and is attached to a pressure-sensitive chamber. As the "fish" sinks, the chamber is squeezed by increasing pressure and the slide moves under the stylus, and is scratched by it. If the temperature also changes with depth, the stylus moves in a different direction and scratches the slide. The combined motion of slide and stylus gives a graph of temperature change with depth.

Among the instruments used to measure the speed and direction of water flow in subsurface currents are water-current meters that provide a direct measure of the velocity of water flow. In addition, some current-meter systems contain a magnetic compass mechanism by which they are able to report the exact direction of water-current flow past the meter. As water flows past the meter, it turns a propeller, or horizontal wheel. These instruments are calibrated so that the number of turns of the current sensor can be converted into a velocity reading. Electrical cables or telemetering equipment attached to the current meter make it possible for current measurements to be received immediately aboard ship (Fig. 3-6).

Buoys are being designed which will be anchored in the deep ocean with sensors and recording apparatus operated by self-contained power units. Long-term characteristics of the water cross-section being sampled will be recorded and transmitted to the surface ship when it makes its periodic call at each buoy.

"Neutrally-buoyant" floats have been designed that can be set to sink to any desired level and float with the current at that depth. Some are attached to surface floats that can be tracked visually, by radio signals, or by ship-board radar. Another type of float, called the *Swallow Float,* emits "pings" that can be picked up by the research ship's acoustical equipment. In this way, both current direction and velocity can be determined.

Another current-measuring device used in conjunction with underwater photography is the *suspended-drop current meter.* This device operates in the following manner: A drop of colored liquid is released in the field of view of a motion picture camera. This drop is small and is composed of a liquid having the same density as the surrounding water. Therefore, its motion, as recorded by the camera, corresponds to the motion of currents found at that depth and location. The motion picture frames can be analyzed to determine both the velocity and direction of the current.

4

The Earth's Structure

Ever since man first began to speculate about his place in the universe, he has constructed theories to account for the formation of the earth. It is not surprising, then, that at least one ancient philosopher theorized that the earth came from the sea. Nor is such a theory quite as absurd as it may at first appear; for it may be that the continents represent huge granitic "blocks" pushed up out of the sea. The origin of the earth's continents and ocean basins remains one of the great unresolved questions of science. But the answer to this and many other related questions will come only when we have learned a great deal more about the internal structure of the earth—the structure and composition of the material that lies below the crust. For that reason, it is worthwhile reviewing what is currently known about the earth's structure and the methods used to study it.

The Origin of the Solar System

The two most popular theories concerning the origin of the earth are the so-called *tidal hypothesis* and the *nebular hypothesis.* In its modern form, the tidal hypothesis pictures another star passing so close to our sun that a huge mass of molten material was pulled from it in the form of a single tidal "filament" of molten material. The planets, including the earth, were created when the filament broke apart.

According to the nebular hypothesis as first proposed, the solar system developed from a diffuse, hot, gaseous nebula. As the nebula cooled it shrank, leaving a series of rings, which condensed into the planets and their satellites. This theory has been modified by Carl von Weiszäcker of Germany and

G. P. Kuiper of the United States. They envision that the rotating disc of dust and gas broke up into eddies of irregular size and arrangement within a radial zone. Each tiny particle of dust, or *planetesimal,* would add its mass to others, and finally there remained separate masses of gas and dust at various distances from the sun. These masses have been referred to as *protoplanets.* Eventually, because of the pressure exerted by the massing of planetesimals, the protoplanet would become molten. After the molten period would come a cooling and solidification of at least the crust of the planet. The planet might retain enough heat to remain completely molten for many millions of years.

The nebular hypothesis is today the generally accepted theory. But, in any case, both the nebular and tidal hypotheses of the earth's origin call for a completely molten state in some early stage of development. What facts about the earth have led scientists to this conclusion? Previously, it had been known that the temperature in mines and oil wells increases with depth. From temperature readings in mines and drill holes, it is estimated that the temperature increase varies between 1° F for every 30-foot descent to 1° F for every 250-foot descent (an average of 1° C for every 43-meter increase in depth).

Another piece of evidence suggesting a past molten state has come from analysis of meteorites that have fallen on earth. All meteorites have been found to belong to one or another of three groups: they are solid nickel-iron or solid rock, or a combination of both. If these meteorites originated from the asteroid belt between Mars and Jupiter and are the remnants of a former planet, then the segregation of planetary materials into concentrations of heavy nickel-iron and relatively lighter rock-silicates needs some explanation. Moreover, the mass of the earth is greater than we would expect it to be, calculating on the basis of the density of crustal rock and a "normal" increase of density with depth (i.e., with increasing pressure). There must be very dense material under the surface rocks. The nickel-iron composition of some meteorites, measured density of the whole earth, and other physical evidence has led scientists to the conclusion that the heavier elements, such as nickel and iron, had migrated toward the center of the earth earlier in its history.

The Interior of the Earth

But the most convincing evidence that the earth is not "solid rock" all the way through is the occurrence of earthquakes and volcanic eruptions. From very earliest times these natural disturbances—which have often been titanic in their destructive power—have engendered fear and much concern about their origins.

Most earthquakes seem to result from the movement of rock along fractures (*faults*) in the earth's crust. Such movements can occur both on land and in the sea bottom. These movements, which are caused by pressure on two adjacent rock areas over a long period of time, can be either vertical or horizontal or a combination of these directions of movement. If the faults or fractures are long (like the 270-mile-long San Andreas fault in California), the tremors can be felt over a large area and they may cause widespread damage.

There are two great "earthquake belts" in the world where the most frequent and noticeable earthquakes have been recorded. One belt circles the Pacific Ocean from Chile to Puget Sound, then through the Aleutian Islands to Japan, the Philippines, Indonesia, New Zealand, and the island groups of the southern Pacific. About 90 percent of the large earthquakes occur in this belt. The second belt extends across the Alpine regions of Europe, the Mediterranean area, including North Africa and a section across Asia (Syria, Jordan, Afghanistan, and India). These active belts are regions where mountains are of recent formation, with steep slopes and many faults.

In volcanic eruptions, man can glimpse at first hand the most dramatic process by which the earth's crustal rocks are changed and built upward above sea level. The Island of Hawaii, for example, has two active volcanoes, Mauna Loa and Kilauea, and the lava eruptions from these and associated volcanoes have actually built the Hawaiian Islands from the floor of the Pacific Ocean.

The distribution of volcanoes in the world does not exactly coincide with the active earthquake belts, but they do overlap in many regions. The majority of the active volcanoes are concentrated in a belt that borders the Pacific Ocean. The volcanoes in this belt are found near the margin of the continents or on island arcs along the continental margin. On the Alaskan Peninsula and the Aleutian Islands are 36 volcanoes that have erupted in historic times. The Kurile Islands and Kamchatka Peninsula have 53 active volcanoes. The Japan arc to the Philippines has 11 active volcanoes. New Guinea, New Hebrides, and the Solomon Islands have numerous active volcanoes, and then the belt continues through New Zealand to the Kermadec Islands, the Tongas, and Samoa. The Hawaiian Islands are an extension of the island trend. On the eastern margin of the Pacific Ocean, no active volcanoes are found from Washington to Mount Lassen in California. The next area with active volcanoes is northern Mexico, and the belt then continues through Central America and the Andes of South America.

In contrast, the continental margins of the Atlantic Ocean are mostly free of active volcanoes and major earthquakes. There are only two regions in the Atlantic that possess active volcanoes: the Lesser Antilles in the Caribbean Sea and the Southern Antilles, which link Patagonia in Argentina with Grahamland in the South Atlantic. Most volcanic activity in the Atlantic Ocean is limited largely to the Mid-Atlantic Ridge (Azores and Iceland) and to the Canary and Cape Verde islands off West Africa.

The second active volcanic belt, the Alpine-Himalayan belt, begins in the Madeira islands, runs through the Mediterranean region to the Aegean volcanoes and continues eastward to Mount Ararat in eastern Turkey and Mount Demavend in Iran. The belt reappears in Burma and continues southward through the Bay of Bengal to Sumatra, (with 12 active volcanoes), Java (with 20), and the Lesser Sunda Islands (also with 20).

Seismology. The great forces that produce earthquakes and volcanic eruptions have built mountains, altered continents, and etched the ocean floor with vast canyons and mountainous ridges. Thus, the study of earthquakes and volcanoes is necessary to an understanding of the nature of the earth's interior. In particular, the realization that earthquake waves could reveal a great deal about the kind of material below the earth's crust has led to the development of a specialized field within physics and geology known as *seismology*. By analyzing the *seismic waves* of earthquakes, seismologists have made striking discoveries about the earth's interior, even though it remains hidden from man's direct view. In 1906, R. D. Oldham established from seismic evidence that the earth possessed a large central *core*. Evidence for Oldham's theory was based on two kinds of seismic waves that travel from the focus of an earthquake: primary (*P*) waves and secondary (*S*) waves. *P* waves travel through both core and mantle; *S* waves travel through the mantle but not through the core. From studies of shock waves in liquids and solids, it is known that liquids do not transmit *S* waves. Oldham was able to distinguish a separation of core and mantle by studying *P* waves. Also he concluded that the core of the earth, or part of the core, must be liquid, because *S* waves were not transmitted through it. He also stated that the core probably was composed of a dense material—liquid nickel-iron.

In 1919, Beno Gutenberg of Germany calculated that the outer boundary of this core lies at a depth of 2900 kilometers (1800 miles) below the surface. This boundary is called the *Wiechert-Gutenberg* discontinuity. In 1963, Miss I. Lehmann determined that the core had two major layers: an *inner core* and an *outer core*. This was revealed in studies of the refraction of certain *P* waves which enter the inner core. Because of the general way seismic waves behaved in passing through the core, it was suggested that the core was liquid nickel-iron. In 1946, K. Bullen concluded that the material in the inner core was under such great pressure that it could not be liquid, even at the high temperatures assumed to exist at the center of the earth. Therefore, the model of the earth's core calls for a liquid outer core and a solid inner core, both composed of nickel-iron.

Important to the discussion of the earth's structure is the work of the Balkan seismologist A. Mohorovičić. Soon after

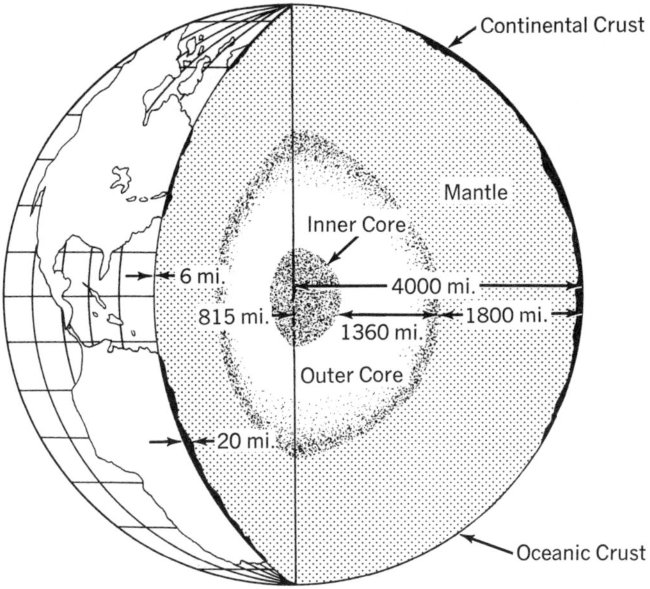

Fig. 4-1. The earth's crust is separated from the underlying mantle by a surface of seismic discontinuity, the Moho. Moho lies at an average depth of 12 km beneath the ocean surface and averages 35 km depth under the land surface.

Oldham's discovery of the large central core, Mohorovičić established from seismic studies a distinct boundary between the earth's surface-crust and the mantle. This boundary has been named the *Mohorovičić discontinuity* in his honor. For the sake of convenience this designation has been shortened to *M-discontinuity*, or *Moho*. The Moho lies at an average depth of 35 kilometers (22 miles) under the continents, but is only about 12 kilometers (7 miles) below the ocean surface. Where the crust of the earth is thicker, the Moho is found at correspondingly greater depths. For example, the Moho lies at a depth of 47 kilometers (29 miles) under the Rocky Mountains. In other instances, it ranges from a depth of 10 to 12 kilometers below the North American Basin to 30 kilometers below the surface in Central Europe.

The picture of the earth's interior that has evolved from the study of seismic records is shown in Figure 4-1. A cross-section of the earth shows concentric layers separated by discontinuities. The crustal layer overlies the mantle layer, which is assumed to consist of a partially molten, dense silicate rock material. Beneath the mantle lies the liquid outer core layer, of nickel-iron composition. At the earth's center is the inner core of solid nickel-iron.

The Continental Crust of the Earth

Seismograph records indicate that 95 percent of the continental volume (the *continental crust*) is made of granitic-type rocks (see Fig. 4-2). The primary rocks of the earth's crust fall into two general categories. One is rich in silicon and aluminum and is called *sial*. The other is poor in these elements, but is rich in magnesium and iron and is called *sima*. The most common sialic rock is granite, the most common simatic rock is basalt. The rocks of the continents are composed mainly of sial. Only sima is found in the rock of the oceanic crust.

Granite is a rock type with which we are familiar. It is a strong, pink-white, or light-gray colored rock that is used as facing stone for lower sections of large buildings. Seismic evi-

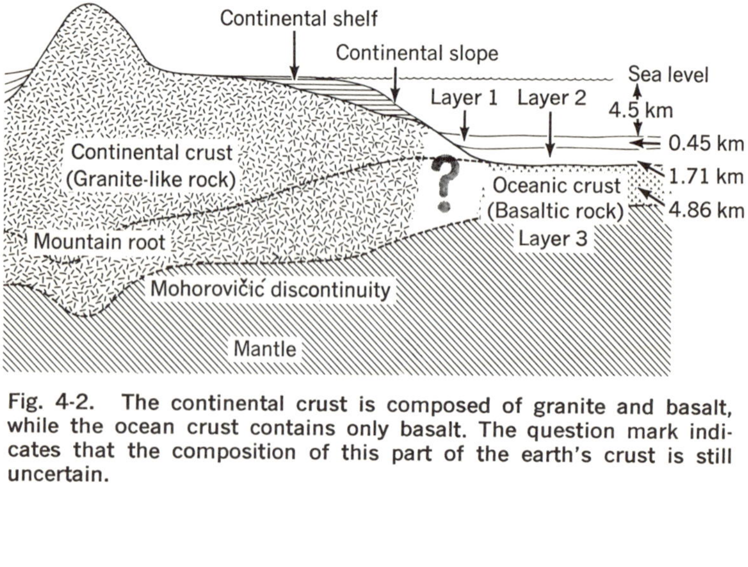

Fig. 4-2. The continental crust is composed of granite and basalt, while the ocean crust contains only basalt. The question mark indicates that the composition of this part of the earth's crust is still uncertain.

dence indicates that the earth's crust under the ocean is composed of basalt. Basalt is a strong, dark gray to black colored rock that frequently appears in volcanic eruptions.

The concept of mountain "roots" composed of continental granite rock lying in a bed of basaltic rock has been one of the most controversial geological questions. Whether the mountain range with its root "floats" in the basaltic rock or whether it resembles a tooth imbedded permanently in the jaw are questions yet to be answered. Many scientists are split on this topic as well as on the topic of granite deficiency in the ocean crust. Some scientists have claimed that the question on granite deficiency should be reversed—Why is there granite in the continents? Their claim is that the oceanic basins are the undisturbed portions of the earth's surface, with an underlying original basaltic crust covered by diverse sediments. The continents are then deformed belts of the earth's surface that have been raised through geologic time.

Soil, sediment, and rocks of sedimentary, metamorphic, and igneous varieties form a thin veneer over the huge granitic masses of the continental crust. Scientists actually know very little about this veneer. Direct observations of rocks lying beneath the surface are possible in mine shafts, but the deepest mine shaft is only 3354 meters (11,000) feet deep. Oil-well drilling has allowed geologists to examine samples of crustal rocks from a maximum depth of 7724 meters (25,335 feet.) Core drilling reveals the layered sequence of sedimentary rock (Fig. 4-3). What information has been collected concerning the continental crust has come largely from seismograph records. The interpretation of seismograph records depends on an educated guess as to how rocks of different compositions would behave at lower levels where pressure and temperature are much higher than can be attained in laboratory studies of rock deformation.

The Oceanic Crust of the Earth

The amount of knowledge concerning the composition of the oceanic crust has come partially from photographs of the sea

Standard Oil Co. (N.J.)

Fig. 4-3. Rock cores obtained from drilling ships are stored for later examination.

floor, dredges of rock and sediment, sediment cores, and moderately deep exploratory oil wells drilled into sediments and sedimentary rocks of the continental shelves. Prior to World War I, deep-sea soundings were made with hemp rope. This method was used in 1840 by James Clark Ross in Antarctic waters to a depth of 4435 meters. Toward the end of the century, piano wire replaced hemp wire, and the regular ship winches were replaced by special sounding winches. A greater part of the scientific knowledge about the oceanic crust has come from seismic shooting at sea. After World War I, echo-sounding systems gave more reliable profile studies than any other method. The first systematic sounding of an ocean was accomplished in the 1925–1927 expedition of the German survey and research vessel *Meteor* in the South Atlantic.

Seismic Reflection Shooting. If natural seismic waves from earthquakes could be used to determine the earth's structure, why couldn't artificial seismic waves from explosions be used to investigate the rock layers of the continental crust or the sediment layers of the ocean floor? Artificial explosions could be limited so that no destruction would result from their generation of seismic waves. Because of the small amount of seismic energy released by such man-made explosions, one of

Gulf Oil Co.

Fig. 4-4. Seismic reflection shooting in the Gulf of Mexico.

the first problems was to design a very sensitive seismic-wave detector, or *geophone*. The geophone could receive the information relayed by ground vibrations from the explosion. Technical advances by companies searching for new oil deposits on land allowed a few pioneering geophysicists to adapt seismic flection shooting to shipboard investigation of the oceanic crust. This method was first employed by Maurice Ewing and others in seismic surveys of the continental shelf begun before World War II.

Seismic reflection shooting requires a single ship that tows a string of geophones (called *hydrophones*) adapted for oceanic seismology. The explosion of a small dynamite charge (Fig. 4-4) in water emits sound waves that penetrate into oceanic sediment and rock. For example, suppose a sound wave strikes a smooth surface and that the path of this sound wave is represented by a ray—that is, by a line perpendicular to the advancing wave front. By the law of reflection, the angle of reflection r for this ray equals the angle of incidence i and the reflected ray lies in the plane of incidence (Fig. 4-5). Part of this sound energy is reflected back to the hydrophones every time the sound waves strike a reflecting "horizon." The reflecting horizon usually consists of sediment layers of different composition or rock layers below the sediment.

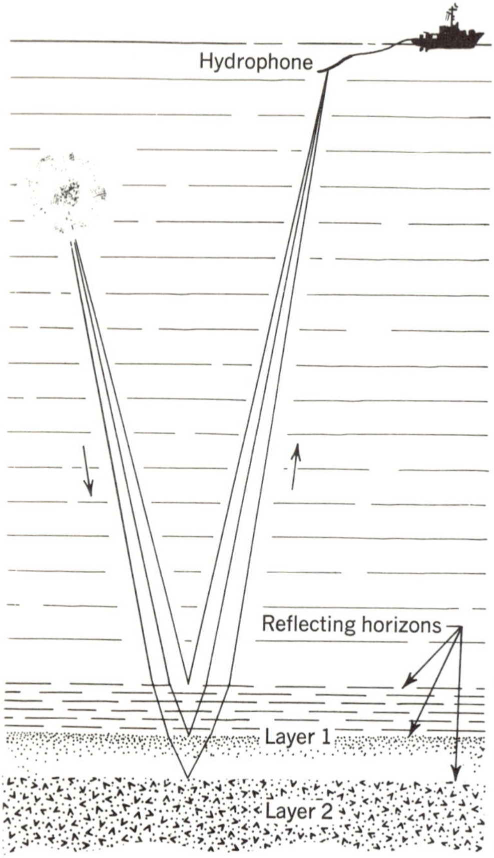

Fig. 4-5. In seismic reflection shooting, a research ship drops a charge at some determined point. Sound waves emitted by the explosion will reflect off different sediment layers. These reflected waves will then be received by the hydrophone system of the ship.

The hydrophone converts the reflected sound waves into electrical impulses that are transmitted to the ship and recorded.

Seismic Refraction Shooting. Seismic refraction shooting requires the use of two vessels: a *shooting ship* that drops explosive charges as it moves along a predetermined course, and a *receiving ship,* which remains stationary and silent so that the faint seismic waves can be detected above the background noise in the ocean. The explosive charge dropped by the shooting ship commonly ranges between a half-pound of TNT to a

300-pound Navy depth charge. The charge, of course, depends on how deep a penetration of sound energy into the oceanic crust is desired and also on the distance between the ships during detonation. As shown in Figure 4-6, seismic waves from the explosion penetrate through the layers of sediment and rock. The velocity of the waves in a solid depends upon the elasticity and the density of the material. The waves are refracted by changes of material or changes of density. Within seconds or less, the refracted waves begin to arrive at the detecting hydrophones. Because of the large size of the explosive charge, the relative thinness of the oceanic crust, and the distance between shooting and receiving ships, it is possible for the seismic waves to penetrate below the Moho to the upper mantle.

Continuous Seismic Profiling. Seismic reflection and refraction techniques are not very useful for the investigation of the thin sediment layers. In order to obtain a detailed record, it became necessary to design a sound source that would operate continuously and emit a controlled-frequency sound. Since regular dynamite explosions generate sound waves of all fre-

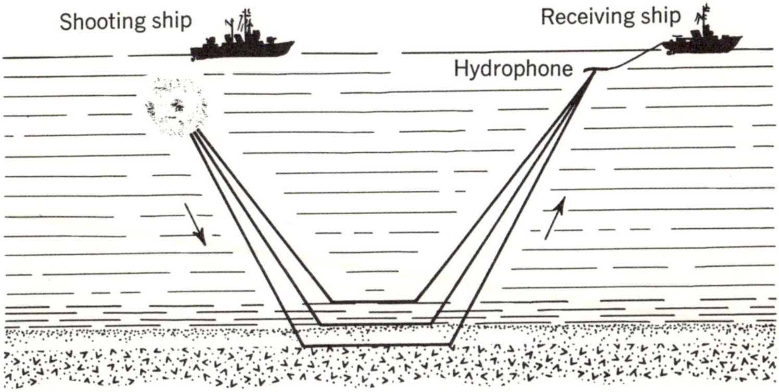

Fig. 4-6. Two research vessels are used in seismic refraction shooting. When a depth charge is released from one vessel at a predetermined point, sound waves will travel along the interface of two sediment layers. The refracted waves will be received by the hydrophone system of the second vessel.

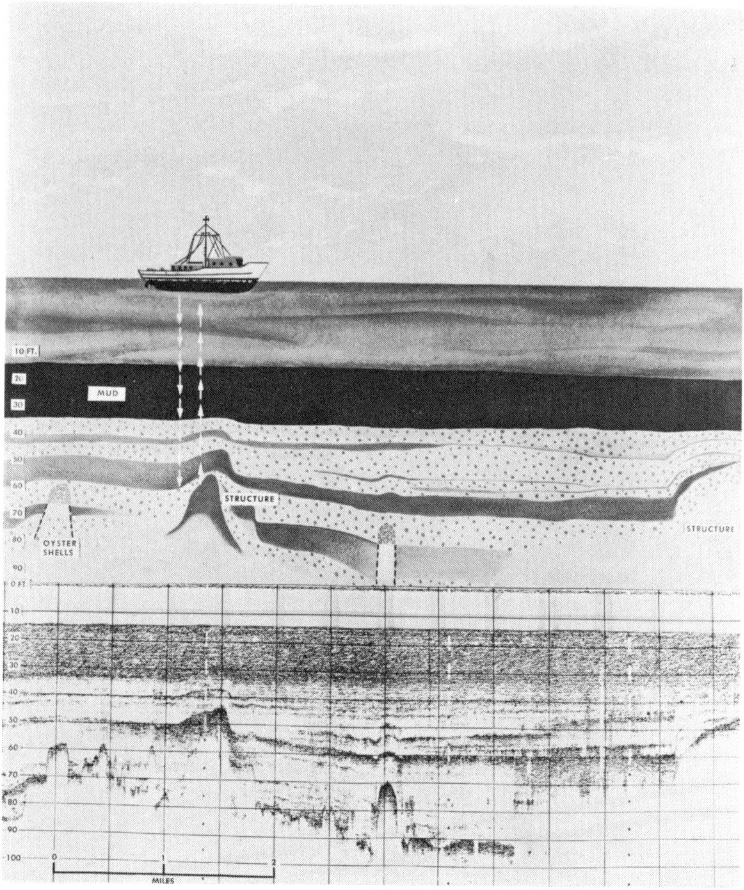

Socony Mobile Oil Company, Inc.

Fig. 4-7. Marine sonoprobe which uses the principles of underwater acoustics is a device that can map shallow sedimentary formations with remarkable accuracy. The marine sonoprobe can penetrate as much as 45 to 90 meters (100 to 200 feet) beneath the sea bottom.

quencies, the structure of the sediment layers had been hidden in the jumble of sound frequencies picked up by the hydrophones. A continuous sequence of controlled-frequency sound waves are obtained by such means as gas explosions (with a *gas exploder*) and electrical discharge (with a *sparker*). The amount of seismic energy released in each sound pulse from a sparker is equivalent to about 0.1 gram of TNT. As a consequence of the small release of energy in each pulse, the penetration depth is currently limited to about 450 meters (1500 feet) below sea level (see Figs. 4-7 and 4-8). Both reflection and refraction techniques are possible with the gas exploder and sparker. It may be possible to design continuous sources that emit a greater amount of energy. With these devices the deep oceanic crust will be opened to a detailed probing that will reveal more information about sedimentary layering.

In comparison with the complex structures found in sedimentary basins on land, salt domes, block mountains, and other masses of the continental crust, the rock structure of the oceanic crust may appear to be rather simple. But, if anything, it is even more complicated.

It is known that the approximate thickness in the low-velocity layer down to the first good reflecting surface in many parts of the Pacific averages about 290 to 330 meters (950 to 1080 feet), and in the Atlantic it averages about 400 to 800 meters (1300 to 2600 feet). The Arctic Basin reflects waves at depths of 250 to 300 meters (820 to 985 feet) below the floor. The margin of ocean basins in some places contain thick masses of

Official U.S. Geological Survey Photograph

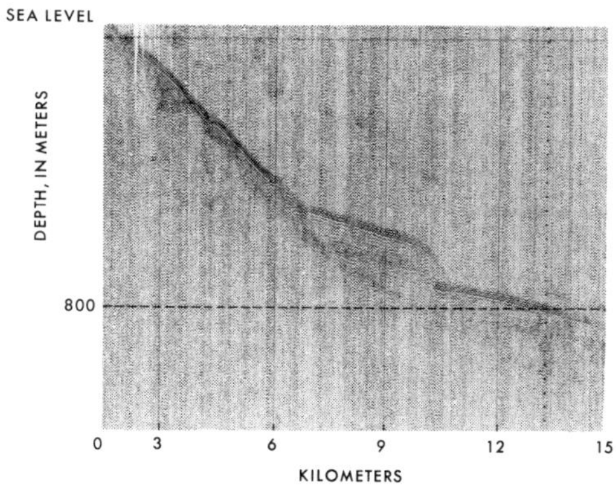

Fig. 4-8. Sedimentary rock layers along the continental shelf and slope are recorded by the continuous seismic profilier. The above profile is the continental shelf and slope southeast of Block Island, Rhode Island.

sediment. A thickness of 5200 meters (17,000 feet) of sediment has been estimated off the east coast of the United States under the continental shelf.

The sediments of the first layer originated from mechanical or chemical destruction of rocks, volcanic disruption, red clay, organic decomposition (planktonic, nektonic, and benthonic forms) which will result in calcareous and siliceous oozes, and minerals which formed when water is oversaturated with soluble material. This sediment is termed *unconsolidated* and *semiconsolidated*. With increasing depth, red clay, and calcareous and siliceous particles undergo various changes. There is a rapid decrease of porosity in clay with increasing pressure; the clay forms a dense mass in a process called *compaction*. Free water is squeezed out with some absorbed water, and plastic deformation and "crusting" of the sediment particles begins. Crystallization, cementation, and recrystallization follow in this pattern, with the resulting formation of low-porosity shales. This is called *lithification*. The so-called *second layer*, which lies under the unconsolidated and semiconsolidated layer is possibly composed of consolidated sediment and lava flows. It has been estimated to be 4900 meters (16,000 feet) thick. Not only clay but also calcareous ooze consolidates, though the lithification rate is lower. This ooze eventually forms a strong, rigid structure, *foraminiferal limestone*. The third layer, the true oceanic crust, is about 6700 meters (22,000 feet) thick and is composed of basaltic rock (Fig. 4-2).

Drilling to the Earth's Mantle

The Mohorovočić Discontinuity, or Moho, lies at an average depth of 12 kilometers (7.0 miles) below the ocean surface. In some areas, it is only 7 kilometers (4.3 miles) beneath the sediment layers. Refraction studies have indicated that the Moho dips rather steeply toward the continent under the oceans. Under the continents, this boundary is found at an average depth of 35 kilometers (21.5 miles). The mysterious mantle lies beneath the Moho. There is much speculation about its composition. Some investigators believe that the

Moho marks the original crust of the earth, whereas others think that the basalt layer in the oceanic crust is the original surface of the earth. Another theory is that Moho represents only a surface of discontinuity in which there is a sharp increase in density of crustal rocks. Alternatively, the Moho may mark the boundary between crustal rocks and a chemically different rock type in the mantle. Since the mantle comprises 84 percent of the earth's interior volume, the composition of the mantle is an important geological question.

With knowledge of the mantle's composition, the petrologist could determine whether the continental crust is only a slag of relatively light weight granitic rocks which collected on the surface of a molten earth soon after its beginning or whether the continents have grown slowly over the geological ages. Furthermore, some answers may be applied to an analysis of the structure of all planets. The question of rising sea level and sinking ocean basins—of when and how the waters have appeared—may be answered through the study of the mantle. Radioactive dating of mantle material may indicate the age of the earth. Questions of the origins of the ocean basins, origin of ocean waters, continental drift, the earth's temperature, causes and prediction of earthquakes, and the beginnings of life may be answered or partially answered.

Project Mohole

In the spring of 1957 a group of distinguished earth scientists, known as the American Miscellaneous Society (AMSOC), suggested a project for sampling the mantle rock in order to answer such questions. Thus, *Project Mohole* was born. It would be an effort to drill a hole to the earth's mantle. Since the earth's mantle is closest to the earth's surface in the deep ocean, drilling a hole to the mantle from a ship seemed the most logical approach to the problem. The novelty and difficulty of this project may become clearer when we realize that the most ambitious drilling program to date had been limited to the relatively shallow water of the continental shelf. The

Socony Mobile Oil Company, Inc.

Fig. 4-9. Offshore drilling platform.

deepest off-shore hole is 20,000 feet. As shown in Figure 4-9, most off-shore oil wells have been drilled from platforms built on pillings sunk into the sediments of the continental shelf.

Fortunately, oil companies had already begun to drill for oil in deeper water. Until the period when man can live and work on the ocean bottom, the only approach to deep water drilling must be from a surface platform or ship. The first problem was to design a ship propulsion system that would allow a drilling ship to remain motionless during the drilling operation. As shown in Figure 4-10, drilling ships such as the *Eureka* contain advanced propulsion systems to hold the ship "dynamically motionless" without anchor. Two 200-horsepower propulsion units are used. Each can be turned through a complete circle, horizontally, to thrust the ship's bow or stern against wind and current. Rolling of the ship cannot be eliminated completely even with the addition of gyroscopic stabilizers. To solve this problem, engineers developed a flexible joint that allows the ship to roll without transmitting tension to the drill pipe. The drilling depth capability of the *Eureka* is 1220 meters (4000 feet) of water.

Shell Oil Co.

Fig. 4-10. The 136-foot-long core-drilling vessel *Eureka* is capable of being held "dynamically motionless" while drilling holes in the sedimentary rocks of the continental shelf.

A larger and more stable ship than those of the *Eureka* class was necessary for Project Mohole. Phase I of the project entailed test-boring in deep water. A modern, 266-foot drilling ship, designed to complete oil wells in oil-bearing formations discovered by smaller drilling ships, had been built by a group of oil companies (Continental Oil, Union Oil, Shell Oil, and Superior Oil). It was named *CUSS I,* using the first letter of each oil company (Fig. 4-11). Willard Bascom, project engineer for Phase I, modified *CUSS I* by placing four 200-horsepower motors around the hull so that the dynamic positioning system would keep the ship motionless in winds up to 37 km per hour (20 knots) and surface currents up to 0.55 km per hour (0.3 knot). This position-stability was necessary to prevent snapping of the 3-km (9800-foot) drill pipe which lay between the ship and the ocean bottom. This system kept *CUSS I* within a ship's length of its desired position during tests off Guadalupe Island near Baja California, Mexico. On April 6, 1961 a test hole was drilled to a depth of 180 meters (590 feet) in a record depth of 3300 meters (10,800 feet) of water. This was a successful operation in that it proved drilling could be accomplished in deep water. It also advanced the

Shell Oil Co.

Fig. 4-11. *CUSS II* is a large drilling vessel currently being used to drill oil wells into the continental shelf. She is the sister vessel of *CUSS I*, which was modified and used to drill the test hole for Project Mohole.

drilling operation of Project Mohole to truly colossal proportions. This operation requires boring through 6000 meters (19,700 feet) of oceanic crust to reach the Moho, while the whole drilling apparatus must be held motionless in water 4500 meters (14,750 feet) deep.

It must be remembered that the deepest drill hole on land, 7724 meters (25,335 feet) down in the West Texas oil field, took two years of drilling to complete. Mohole drilling will probably require a longer time and will necessitate a platform that can be held motionless for a period of years. The platform will float on two submersible submarine hulls, each 390 feet long and 35 feet in diameter. The three-deck platform will be positioned 88 feet above the hulls on six columns, each 31 feet in diameter (Fig. 4-12). The upper deck will be 279 feet long, 234 feet wide, and 23 feet high. The structure will be submerged to 70 feet for stability and its displacement will be

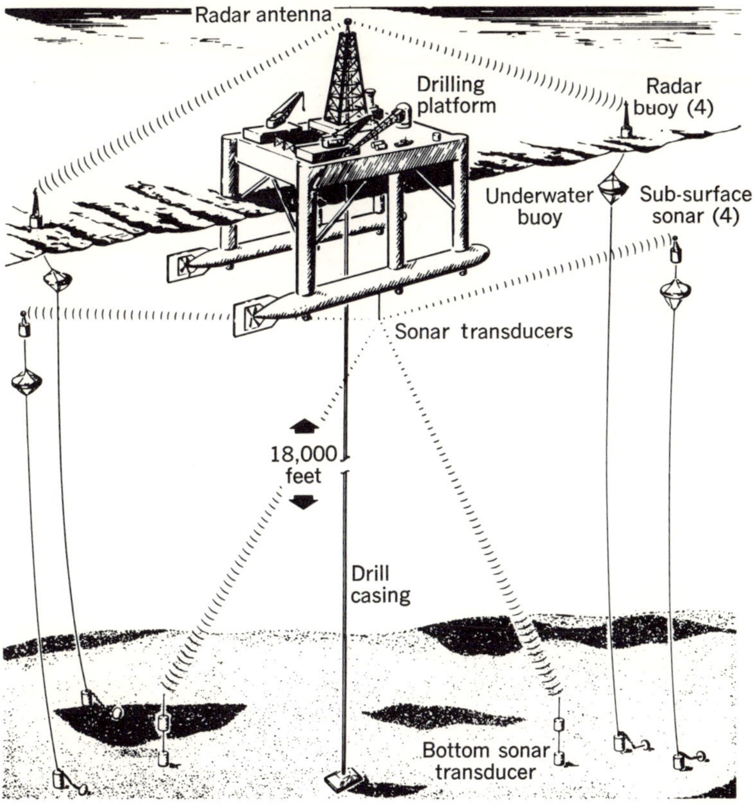

Honeywell

Fig. 4-12. The site for Project Mohole lies about 100 miles north of Maui in the Hawaiian Islands—an area where explosion soundings have shown that the mantle rises to within three miles of the ocean floor, beneath two and a half miles of water.

24,000 tons. Positioning units below each of the six columns will be rated at a maximum of 550 horsepower. In addition to the six positioning units, there are two main propulsion units which give the structure a cruising speed of 8-10 knots. To maintain the structure's position within a 1000-foot circle there will be a sonar array either at the ocean bottom or submerged at shallow depths. Since the drill will be going through a few thousand feet of sediment, the structure will possess a mud pump system which can pump a volume of 550 gallons per minute. These were some of the surface engineering problems that scientists faced during the entire planning of Phase II.

The underwater problems were another factor, and a new factor. Since this would be the first opportunity to bore into unknown layers and at tremendous depths, scientists were concerned about temperature and pressure on the drill pipe, the scientific instruments, the drill kit, and the power to drill. They faced the problem of placing the drill string back into the hole at the ocean bottom after the pipe had been withdrawn for inspection or to replace a worn bit. The suggestions to cope with this problem have been promising: 1) the use of a riser pipe from the operating deck to ocean bottom, so that the hole could be always located; 2) the use of sonar direction finding. As for the drill pipe itself, it must support its own weight and withstand the rigors of rotation, bending, and drilling during the operation. The most promising material is a titanium alloy that is highly resistant to corrosion and metal fatigue. Scientists have developed prototype instruments for "down-hole" measurements that appear to be resistant to high temperatures and pressure. An improvement of the turbodrill will not require rotation of the drill pipe. The prototype turbine will have 22,000 horsepower. With these problems solved, Project Mohole will begin actual drilling in late 1965 or 1966. In the meantime Russian geologists are reported to be looking for suitable locations from which a land-based drill rig could penetrate to the mantle, thus avoiding some of the problems associated with the United States project.

The discussion thus far has outlined briefly the structure of the earth, the methods used to investigate its structure and surface topography, and the significance of such studies. In the next chapter we shall see what these studies have revealed about the topography of the ocean floor.

5

Topography of Ocean Basins

Explorers and traders criss-crossed the major oceans for over three hundred years without making an effort to chart the topography over which they traveled. The only depth-measuring device that they possessed, the *sounding lead,* was used in shallow coastal waters. Many ships have passed safely over a sand bar or through a passage in a reef with the aid of lead and hemp-line soundings. Channel buoys and dredged channels eased the coastal pilot's job of maneuvering ships into the numerous world ports that are surrounded by dangerously shallow water. Since these sailors had no navigational need for a knowledge of water depth in the open ocean, the topographic panorama of the ocean bottom remained unknown to man. It was just assumed that the ocean was deepest in the middle, somewhat in the nature of a saucer or soup dish. It came as a surprise when soundings revealed that the deepest parts of the ocean were to be found relatively close to land. In general, this relationship holds true for all ocean basins.

Bathymetric Charts

The first successful deep-sea soundings were made by James Clark Ross with hemp line in 1840. He sounded a depth of 4800 meters (15,750 feet) in the South Atlantic. By 1854 enough reliable soundings had been made so that Matthew Maury was able to draw a preliminary *bathymetric chart* of the southern part of the North Atlantic Ocean. A bathymetric chart indicates ocean depth by use of lines called *isobaths,* which are drawn through all points in the ocean basins that are the same depth below sea level. To illustrate, suppose we were to fill a soup bowl with water. If a portion of the water were removed, we would draw a line around the inside of the soup bowl to show the position of the new water level. We would then repeat the procedure by removing enough water to cause an equal drop in the height of the water level. We would continue this process until we reached the bottom of the bowl. If we refilled the bowl with clear water, we would see the lines marking equal increments of depth below the water surface. These lines are similar to the isobaths found on a bathymetric chart. However, there are some obvious differences between the two sets of lines. The simulated isobaths in the soup bowl appear as smooth concentric circles. But in the oceans the isobaths must follow—horizontally—the *contour* of the ocean bottom topography, which is much less regular than the sides of the soup bowl. Isobaths are sometimes called *underwater contour lines,* because they show contour of the ocean basin. Thus, on a bathymetric chart, each depth contour has many crenulations. Each crenulation represents some irregularity in the topography found at that depth level.

One of the methods in which the bathymetric chart can be used is indicated in Table 5-1, which lists the surface area of the oceans seaward of the continental shelves. Figure 5-6 illustrates a bathymetric chart of the area of the west-central Pacific Ocean. The bathymetric chart (see Table 5-1) shows that the surface area of deep water amounts to slightly over 335 million square kilometers. It must be remembered that 362 million square kilometers of the earth is covered by water.

Table 5-1. Surface area of the oceans seaward of the continental shelves.

Ocean	Area (Square Kilometers)	Area (Square Miles)
Pacific	165,245,000	63,801,000
Atlantic	82,440,000	31,830,000
Indian	73,442,000	28,356,000
Arctic	14,090,000	5,440,000
	335,217,000	127,427,000

We can see from this chart that nearly 92 percent of the water-covered earth is contained within the boundaries of the deep ocean basins. The table also shows that the deep Pacific Ocean has a surface area approximately equal to the combined area of the other deep-ocean basins.

The distribution of depth within the ocean basins proves to be very interesting. A bar-graph of the percentage of ocean area at one-kilometer intervals of depth below the surface is shown in Figure 5-1. It is shown that 33 percent of the oceanic area is found between 4–5 kilometers (13,000–16,400 feet) below the surface. No other depth range includes as much area. Four-fifths of the water-covered earth contains depths of 3300 meters (10,800 feet). The continental shelves compose three percent of the water-covered earth. From information contained in Table 5-2, scientists have estimated that the *average* depth of the oceans is 3.8 kilometers (12,500 feet). In particular, the Atlantic Ocean has a mean depth of 3870 meters (12,700 feet), the Indian Ocean of 3960 meters (13,000

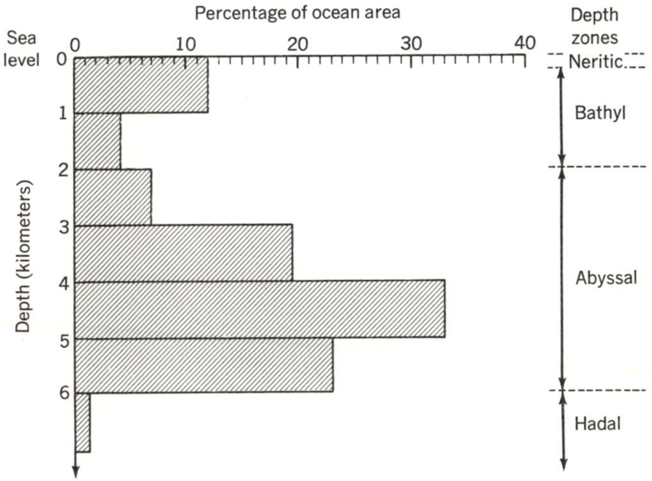

Fig. 5-1. Bar-graph showing the percentage of ocean area found at various depth levels.

Table 5-2. Deep-Sea Trenches.

Ocean	Depth (meters)	Depth (feet)
Pacific Ocean		
Mariana Trench (Challenger Deep)	10,915	35,800
Tonga Trench	10,853	35,597
Kurile-Kamchatka Trench	10,543	34,580
Philippine Trench (Mindanao Trench)	10,033	32,907
Kermadec Trench	10,003	32,809
Japan Trench (Idzu-Bonin Trench)	9800	32,153
North Solomons Trench	9142	29,987
New Hebrides Trench	9038	29,643
Yap Trench (West Caroline Trench)	8602	28,216
Palau Trench	8142	26,706
Peru-Chile Trench	8057	26,427
Aleutian Trench	7672	25,165
Nansei Shoto Trench (Ryukyu Trench)	7512	24,639
Middle America Trench	6662	21,851
Atlantic Ocean		
Puerto Rico Trench	9392	30,184
South Sandwich Trench	8262	27,100
Indian Ocean		
Sunda Trench (Java Trench)	7252	24,442

feet), and the Pacific Ocean of 4780 meters (15,700 feet). The oceans basins contain a total of about 300 million cubic miles of water. If the continents and ocean basins were flattened to an equal height, the earth's smoothed surface would be covered by 2430 meters (8000 feet) of water.

Depth Zones

One of the objects of this chapter will be the examination of the relationship between topography and depth. For convenience in describing oceanic depth, a verbal classification of depth zones is given on the right-hand side of Figure 5-1. The continental shelf environment, roughly from the sea surface to a depth of 0.2 kilometer (650 feet), is referred to as the *neritic zone*. The *bathyal* depth zone lies between 0.2 and

2 kilometers (650 to 6500 feet). *Abyssal* depths range between 2 and 6 kilometers (6500 to 19,700 feet) below the ocean surface. The trench environment lying below 6 kilometers is termed the *hadal* depth zone. In terms of percentages, it should be noted that 82 percent of the ocean bottom is found within the abyssal depth zone—making this depth zone the most important one in the study of ocean-basin topography and physical processes.

Depth Sounding

The first closely spaced, systematic soundings were taken in conjunction with the laying of the first transatlantic cable between Newfoundland and Ireland in 1858. Prior to the mid-1800's there were many problems that had made bathymetric surveys impractical. One of the problems was obtaining enough hemp line of sufficient strength to carry the sounding lead to the bottom and back. It was also difficult to estimate when the weight had struck bottom, but the oceanographer soon learned to judge when bottom had been reached by feeling a change of tension in the line. If either the ship drifted while sounding was in progress or if undercurrents deflected the line from the vertical position, then the depth recorded by the length of the line would be too great. If the angle between the sea surface and the sounding line could be measured, simple trigonometry would permit calculation of the actual depth.

Some improvement in sounding procedure resulted from the substitution of hydrographic wire for hemp line. An important breakthrough in determining depths occurred after World War I, when it became technically possible to use sound waves in place of hydrographic wire. In principle, *sonar*, the *echo sounder*, and the *precision depth recorder* use a "pinger" that releases sound energy below the water surface. A small portion of this sound energy strikes the bottom, and an even smaller portion is reflected back to a *hydrophone*. The hydrophone is a sensitive microphone which converts the reflected energy into electrical impulses that are fed into a ship-board depth indicator. With the knowledge of the sound-wave velocity in water and the length of time between the ping and echo, the oceanographer can calculate the water depth. The modern depth sounder employs a strip-chart recorder on which a continuous graph of ocean-bottom topography is drawn while the ship speeds along its course. This particular type of topographic chart is called a *topographic profile* (see Fig. 5-2).

The topographic profile must be corrected for differences in sound velocity that result from differences in water density and temperature. It must also be corrected for multiple echoes from the rugged ocean bottom. The constant improvements in these recorders allow the oceanographer to "observe" finer details of bottom *relief*. *Topographic relief* refers to the difference in elevation between the highest rise and lowest depression within a given area of the earth's surface. For example, Mount Everest rises 8850 meters (29,002 feet) above sea level. The bottom of the Challenger Deep of the Mariana Trench lies 10,915 meters (35,800 feet) below sea level. Therefore, the relief of the entire surface of the earth is 19,765 meters (64,802 feet). When we consider that the earth's radius is 6,378,000 meters, then the topographic relief of the earth is of minor magnitude. It amounts to only 0.33 percent of the length of the earth's radius. However, the human inhabitants

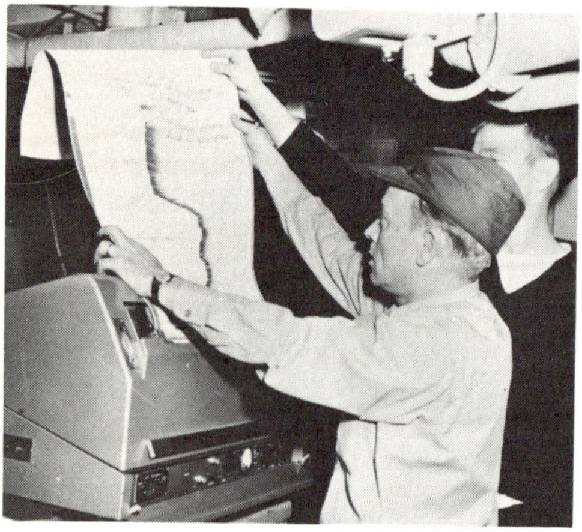

Official U.S. Navy Photograph

Fig. 5-2. Bottom-profile readout from a precision depth recorder.

of the earth can translate this small percentage into a visual appreciation of the surrounding topographic beauty. The snow-covered crest of Mount Everest has beckoned man to climb its walls and to reach the "roof of the world." Perseverence and ingenuity have been basic factors in man's conquest of this lofty challenge. It is only recently that the imagination and engineering ability of a few individuals have allowed man to triumph over the cold constricting darkness of the Challenger Deep. The hazards involved in these conquests have indicated that such trips will never be commonplace for laymen.

Topographic Divisions

We can substitute the topographic trace on the strip-chart recorder for a more descriptive observation of ocean-bottom topography. Continents and ocean basins are classified as the first-order relief features of the globe. As we all know, the continental topography consists of isolated mountains, mountain ranges, valleys, valley systems, plains, and plateaus. The main topographic divisions of the ocean basins include the *continental margin, ocean basin floor,* and *mid-ocean ridge.* Oceanic topography (with the notable exception of the North Pacific Ocean basin) consists of a mid-ocean ridge flanked by the deep ocean basins on both sides. Continental-margin topography begins at the landward sides of the ocean basin floor and ends at the shoreline. Figure 5-3 illustrates examples of continental margin topography.

The Continental Margin. A general and simpler scheme for classification of ocean-basin topography is illustrated in Figure 5-4. The *continental shelf, continental slope,* and *continental rise* are physical features within the continental margin. The relatively shallow, seaward-sloping surface off the continents is the continental shelf. In some areas of the world there may be no continental shelf. In other areas it may vary in width from only a few kilometers up to 650 kilometers (400 miles). The continental shelf ranges in width from 15 kilometers (9.3 miles) off Florida to 500 kilometers (310 miles) a.

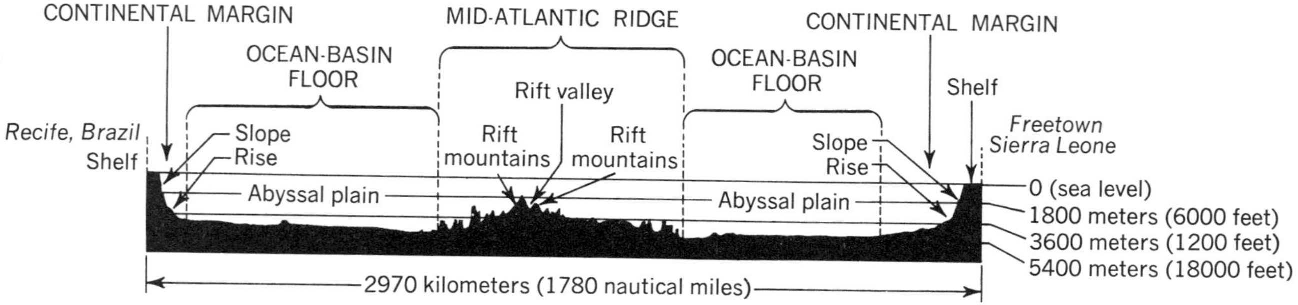

b.

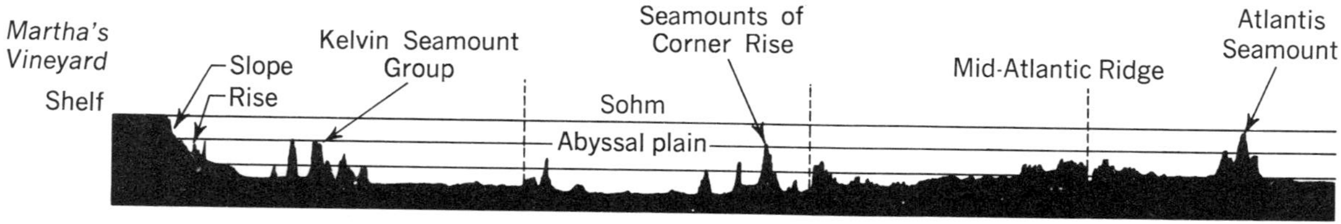

Fig. 5-3. (a) Topographic profile across the Atlantic Ocean, near the equator from Recife, Brazil to Freetown, Sierra Leone. (b) Profile from Martha's Vineyard, Massachusetts, southeast across the Mid-Atlantic Ridge. The Kelvin and Atlantis seamount groups can be seen in relation to the surrounding Sohm Abyssal Plain.

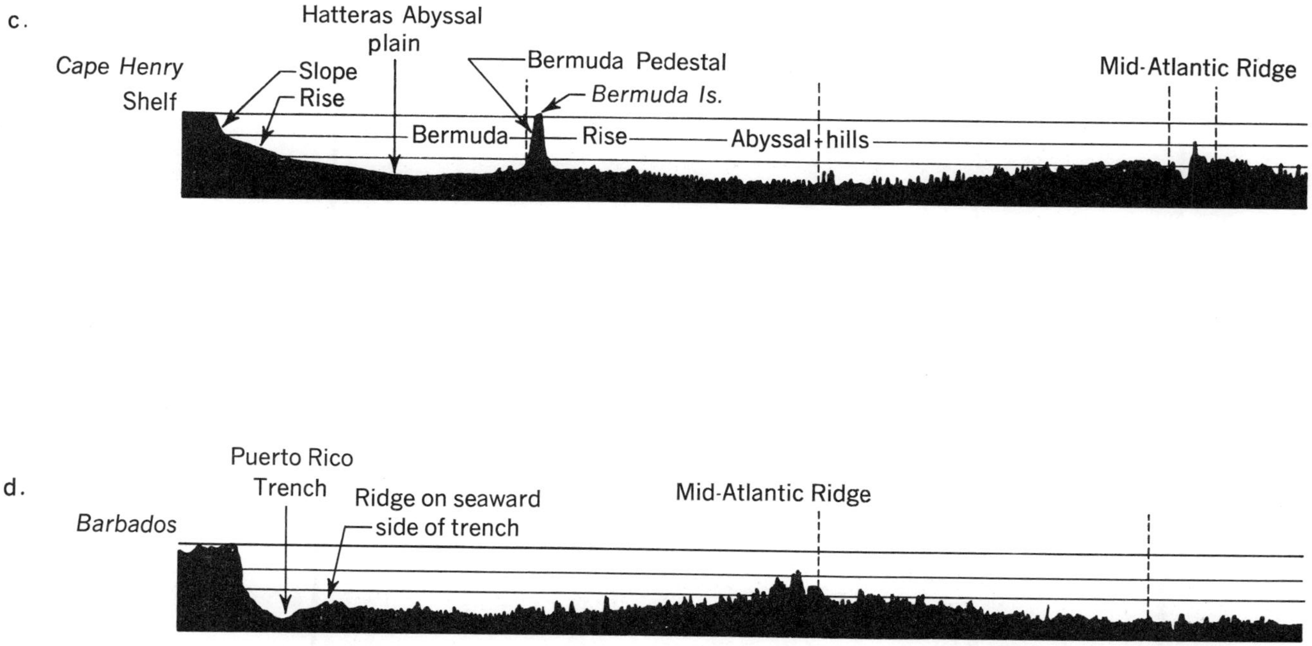

Fig. 5-3. (c) This profile from Cape Henry, Virginia to the Mid-Atlantic Ridge shows the relationship of the Hatteras Abyssal Plain with the Bermuda Pedestal. (d) This topographic profile shows the ocean basin from Barbados in the Lesser Antilles (West Indies) north across the Puerto Rico Trench then east across the Mid-Atlantic Ridge (Fig. 5-3 has been adapted from Heezen, Tharp, and Ewing. Special Paper 65. Published by the Geological Society of America © Copyright 1957).

off the Gulf of Maine; along the east coast of the United States; the shelf averages 170 kilometers (105 miles) in width and encompasses an area of 510,000 square kilometers (196,000 square miles). Off Florida in the Gulf of Mexico, the shelf extends 760 kilometers (470 miles) seaward and then drops one mile within a two-mile horizontal distance. This physical feature would compete with the Grand Canyon or the mountains in Yosemite Park for sheer grandeur and breathtaking heights. The largest shelf area in the world lies off Russia in the Arctic Ocean where it reaches a width of 800 miles. The Pacific coasts of China and Russia encompass a shelf area of one million square miles. Except for these broad shelf areas, the average width of the earth's continental shelves is 42 miles with an average depth of about 180 to 210 feet and an average seaward slope of 10 feet in a mile.

What has been more interesting in oceanic research in recent years is the classification of continental shelves. There are shelves that border glacial land masses and have depths of more than 100 fathoms (600 feet). They extend close to shore

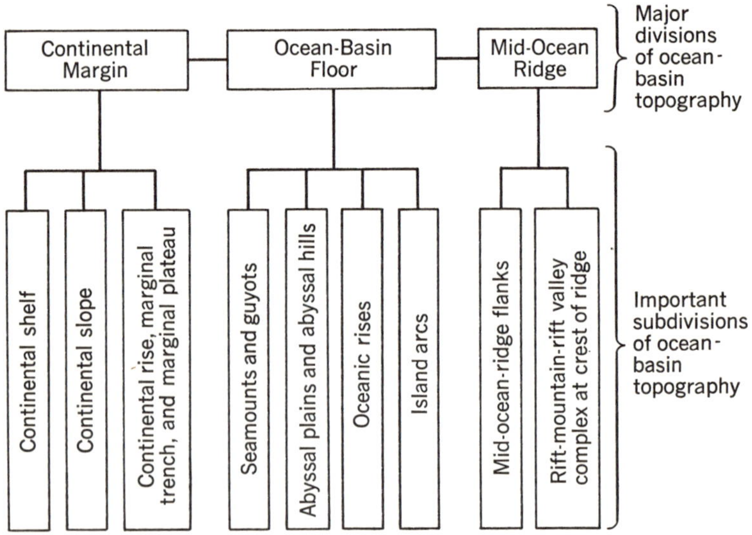

Fig. 5-4. A simplified classification of ocean basin topography.

and have many banks that rise close to sea level. These banks are the best fishing grounds in the world because light penetrates the water over the whole area of the banks. Georges Bank off the New England coast, Banquereau off Nova Scotia, and Grand Banks off Newfoundland are the best examples of this type of shelf.

Along the coastal plains of New Jersey and the area of the North Sea north of the English Channel, elongated sand banks slope seaward with depressions pitted about the banks. Some large river deltas—Mississippi, Indus, Niger, and Nile—have large shelf areas at their periphery. The Great Barrier Reef of northeastern Australia that covers hundreds of square miles has been built up by shallow-water corals. This reef area has been a danger ground for navigation since man first sailed in those waters. Discontinuous valley systems have been found on some shelves, especially the Sunda shelf in the Sunda Sea near Australia. These are shallow branching valleys with a leaf-vein pattern characteristic of many continental river systems.

As the shelf descends slowly seaward, there is an abrupt change in the bottom slope, marking the transition from continental shelf to continental slope. This transition area is termed the *shelf break*. It occurs at a depth of 50 meters (165 feet) off the southern tip of Florida, but is found at a depth of 120 meters (395 feet) off New England. In other areas along the coast line, the shelf break is found at depths as great as 460 meters (1500 feet).

The continental slope extends seaward from the shelf break and generally reaches depths of 1370 to 3200 meters (4500 to 10,500 feet). A continental rise, marginal plateau, or marginal trench may lie at the base of the continental slope. Generally a continental rise is located at the base of the continental slopes in the Atlantic, Indian, Arctic, and Antarctic ocean basins and also in the Mediterranean Sea. Figure 5-3 illustrates continental rises in three different regions.

Submarine Canyons. Frequently along the continental shelves and continental slopes of some areas, oceanographers have found deep canyons beginning at some part of the shelf

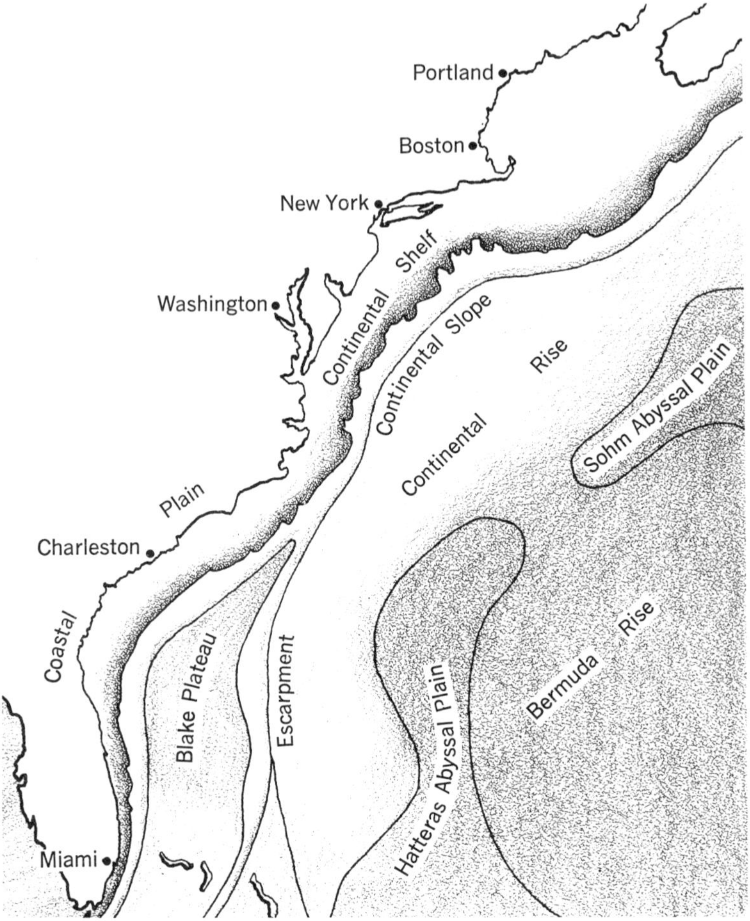

Fig. 5-5. Ocean topography found adjacent to the U.S. east coast.

and opening into the ocean floor. A large number of these canyons have been charted along the northeastern coast of the United States (see Fig. 5-5). Other canyons are located off the Congo river, off southern California, and off the west coast of Corsica in which every bay has a continuation under the sea. Some submarine canyons have been found in the ocean-basin floor. They do not appear to have any land connection. One of the first such *mid-ocean canyons* to be discovered is located in the western North Atlantic, near Greenland. It has an average width of 2 to 4 miles and topographic relief of 50 to 200 meters (165 to 655 feet) along its 2000-mile length.

The origin of these canyons has been a highly controversial subject among oceanographers. In the 1930's R. A. Daly proposed a theory according to which turbidity currents were responsible for erosion of the sea bottom. From Daly's findings and his own model studies, P. Kuenen in 1951 has claimed there are substantial turbidity currents that could cause such canyons. Maurice Ewing and Bruce Heezen (Lamont Geological Observatory) favored the turbidity-current theory on the basis of data from certain earthquakes. Francis P. Shepard of Scripps Institution of Oceanography has proposed an alternate theory. Shepard says that canyons may be products of ancient river cutting at a time of coastal elevation. Slow subsidence of continental margins allowed preservation of the canyons over long ages. The submerged heads of the canyons must have worked into the inner slopes by landslides, while the valleys became extended on the outer slopes as a result of turbidity currents. Although the controversy has not been settled conclusively, the weight of the evidence favors the turbidity-current theory (see Chapter 6).

Marginal Plateaus. Beyond the shelf edge, a major marginal plateau, the Blake Plateau, begins south of Cape Hatteras, North Carolina and extends southward nearly to Puerto Rico. Figures 5-3 and 5-5 picture the relation of this plateau to the other profiles of the continental margin. At its widest portion, the Blake Plateau extends seaward about 275 kilometers (170 miles) from the base of the continental slope at a depth of approximately 610 meters (2000 feet) to a seaward depth of 1000 meters (3300 feet). At about the 1000-meter depth the continental slope begins again, but is very steep and is termed the "Blake escarpment." Core samplings of this plateau indicate a rocky bottom or a bottom covered by a hard calcareous material with little or no soft sediment.

Marginal (Deep-Sea) Trenches. In some areas of the world the continental slopes plunge into deep marginal trenches. The descent into these deep-sea trenches seems unending and terrifying because of the surrounding darkness. They are longitudinal depressions with depths generally in excess of 6000 meters (19,700 feet). Most of them are found in the Pacific Ocean. Table 5-2 lists the trenches of the Atlantic, Pacific, and Indian Oceans with their depths. Only two major trenches have been discovered in the Atlantic Ocean. They are the Puerto Rico Trench and the South Sandwich Trench. The Puerto Rico Trench, the only marginal trench in the North Atlantic Ocean, begins north of Puerto Rico and swings in a seaward-convex arc to the east and south parallel to the islands of the Lesser Antilles. A topographic profile of the continental margin across the Puerto Rico Trench is shown in Figure 5-3. As for the other Atlantic trench, the South Sandwich Trench lies in an east-facing convex arc between Cape Horn at the southern tip of South America and the mainland of Antarctica. On May 10, 1964 the first manned descent to the bottom of the Puerto Rico Trench was accomplished by the French-owned bathyscaph *Archimede*. This vessel is a 21-meter (70-foot) long and 8-meter (26½-foot) high submersible with a personnel compartment for 3 men. It has a similar design to U.S. Navy bathyscaph *Trieste*. The descents into the Puerto Rico Trench were part of a joint French–United States research effort known as "Operation Deepscan."

Marginal trenches generally drop at least 1800 meters (6000 feet) below the level of the ocean floor and are usually separated from the floor by a low ridge that runs parallel to the trench on its seaward side (Fig. 5-3d).

Fifteen major trenches have been discovered ringing the Pacific Ocean basin. The trench between New Zealand and Samoa is sometimes referred to as the Tonga-Kermadec Trench or as two separate trenches, the Tonga Trench and Kermadec Trench. The Mariana Trench, by virtue of having the greatest depth of any trench (the Challenger Deep), has long held the extreme interest of the oceanographic community. The general geographic setting of this trench in the west-central Pacific is shown in Figure 5-6. The eastward-facing convex arc of the trench lies south of the southern limb of the Japan Trench. The arcs of the Yap Trench (also known as the West

Caroline Trench) and the Palau Trench lie southwest of the Mariana Trench. To the west, an almost linear trench lies seaward of the southern islands in the Philippine group; this is the Philippine Trench, or Mindanao Trench. The profile of trenches is illustrated in Figure 5-6. A bathymetric chart with additional "altitude tint" symbols was prepared to illustrate the complexity of the trench in relation to the surrounding topography. Figure 5-7 shows the arc of the Mariana Trench bounded on the east by isolated submarine rises that parallel the trench. To the west, two submarine ridges, the Mariana Ridge and the South Honshu Ridge, have a more broad seaward convex direction. The Mariana Islands rise to the ocean surface as the only visible evidence of the ridge beneath the

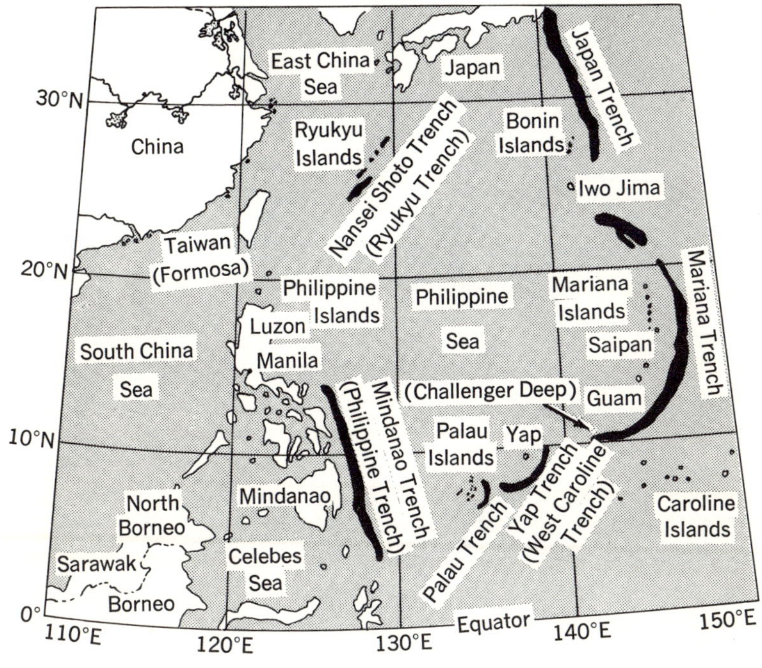

Fig. 5-6. Map of the deepest recorded trenches of central western Pacific Ocean (drawn roughly according to outlines of 3500 fathom contour).

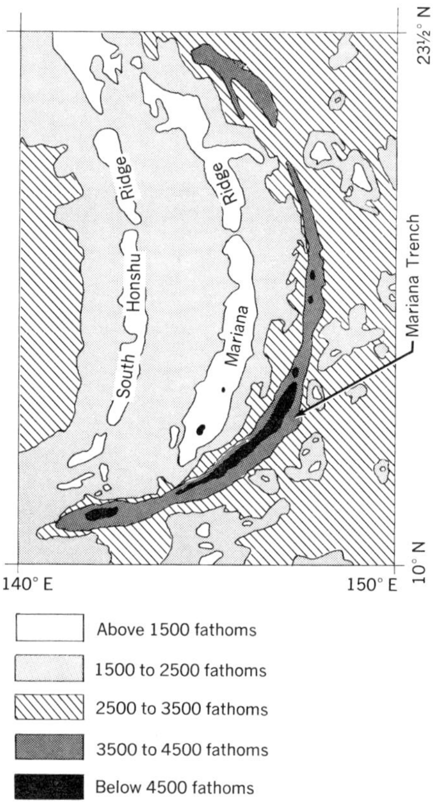

(based on U.S. Navy Oceanographic Office Chart)

Fig. 5-7. Detailed bathmetric chart of the Mariana Islands and the associated Mariana Trench area of central western Pacific Ocean.

surface. Five separate areas within the Mariana Trench have depths greater than 4500 fathoms (27,000 feet or 8235 meters). The southernmost of these areas is the deepest in the world and contains the Challenger Deep into which the bathyscaph *Trieste* descended.

Pacific Ocean trenches range in depths between a shallow 6662 meters (21,851 feet) for the Middle America Trench off the west coast of Mexico to the great depth of the Challenger Deep (10,915 meters). Some of the trenches in the Pacific and Atlantic Oceans have walls steeper than parts of the steep-walled Grand Canyon, which is one of the most impressive chasms on land. Whether a person looks over the edge of the Grand Canyon or the highest local point, he can't believe that such heights exist, and he is disturbed to be so near the edge. But imagine this person on the west coast of Chile in the Andes Mountains approximately 22,500 feet above sea level. If the sea below him were to be drained, he would notice a deep trench, the Peru-Chile Trench, that runs parallel to the coast line. There is a slight continental shelf and slope, and then a shear drop begins. From his observation position, he would be looking down a 22,500-foot mountain rise that at sea level would plunge more than 26,000 feet within a horizontal distance of 100 miles. This is approximately a 9-mile plunge! There can be no doubt that this chasm would be more terrifying than the Grand Canyon or any physical feature of the continents.

The Ocean-basin Floor

Island Arcs. The seaward-convex curve of some trenches lies along the seaward margin of similarly curved island groups. The island groups are called *island arcs,* and the trenches that bound them are sometimes called *island arc trenches.* Island arcs are considered part of the ocean basin floor because they appear to be built on top of typical oceanic crust rather than on the thicker crust of the bordering continental masses. No major differences have been discovered to justify a separation of island arc trenches from their counterparts, the marginal trenches that lie seaward of continental masses. Major earthquakes occur towards the land away from the trenches, and many active volcanoes are found in the island arcs and continents near the trenches. A ring of active volcanoes, a ring of fire, literally surrounds the Pacific.

Abyssal Plains and Abyssal Hills. About one-third of the Atlantic and Indian Ocean basins and three-fourths of the Pacific Ocean basin is considered to be the *ocean basin floor.* Some of these basins are flat and uniform with some bottom configurations. *Abyssal plains* and *abyssal hills* are the most common topographic features found on the ocean basin floor. Abyssal plains (illustrated in Fig. 5-3, and in the physiographic diagram of Fig. 5-8) are the flattest portions of the earth's surface. They have been found, in all oceans, characteristically near the base of the continental rise. The abyssal plains are common in the Atlantic and Indian Oceans, but are relatively rare in the Pacific Ocean basin.

Abyssal hills are small oval-shaped hills that lie seaward of the abyssal plains, except in the Pacific Ocean basin where abyssal hills are found immediately adjacent to the continental margin. Maximum relief of the abyssal hills is approximately 600 meters (1970 feet). Most oceanographers believe the abyssal plains were formed from masses of continental sediment that have buried the abyssal-hill topography.

Oceanic Rises. Large, isolated, non-seismic areas that rise above the abyssal floor are called *oceanic rises.* They form neither part of the continental margin nor part of the mid-oceanic ridges. Although they exist in all oceans, the best example of an oceanic rise is the Bermuda Rise of the North Atlantic Ocean Basin. This rise is shown in the topographic profile of Figure 5-3c and in the physiographic diagram of Figure 5-8. The Bermuda Rise is an oval asymmetrical arch 480 kilometers (300 miles) wide and 960 kilometers (600 miles) long. It lies seaward of the Hatteras Abyssal Plain and is bounded on the west by abyssal hills. A relatively narrow pedestal 80 kilometers (50 miles) wide and 129 kilometers (80 miles) long at the base rises from a depth of 4550 meters (14,900 feet) at the crest of the Bermuda Rise. The top of this pedestal reaches to within 37 meters (120 feet) of the ocean surface. Coral growth from the top of this pedestal has resulted in the formation of Bermuda and associated shoals.

Seamounts and Guyots. A *seamount* is a submerged mountain peak that rises from the ocean floor. Seamounts are considered to have sharp conical peaks, whereas flat-topped seamounts that are found in some ocean basins are called *guyots,* named after the Swiss-American geologist Arnold Guyot. The tops of seamounts and guyots generally lie 1000 to 1460 meters

(3300 to 4800 feet) below the ocean surface. Five hundred of these mountains have been found, since 1941, in the North and Central Pacific basins as well as in the Mariana basin. Only a relatively few seamounts and guyots have been located in the Atlantic.

Guyots are limited in extent and occur along three physiographic lines: (1) the Emperor Seamounts south of Kamchatka, (2) Marcus-Necker Rise west of the Hawaiian Islands, and (3) the other in a zone extending from the Marianas to the Marshall Islands. Ten guyots have been found close together in the Gulf of Alaska.

In the physiographic diagram of Figure 5-8 a group of seamounts, called the Muir Seamount Group, is shown just north of the Bermuda pedestal on the Bermuda Rise. Another group of conical seamounts range in a line from the northern part of the Bermuda Rise, across the Sohm Abyssal Plain, up the continental rise, and into the continental slope near Georges Bank. Some of these peaks are flat-topped. They are illustrated by the topographic profile in Figure 5-3. Another group of seamounts tops the Corner Rise that bounds the Sohm Abyssal Plain on the east. The Atlantis Seamount (Figs. 5-3 and 5-9b) is seen to be only one member of an impressive seamount group that includes Plato Seamount, Cruiser Seamount, and Great Meteor Seamount. This group rises from the east flank of the Mid-Atlantic Ridge south of the Azores Plateau. Seamounts and guyots can be found in the ocean-basin, mid-ocean ridge, and continental-margin divisions.

Seamounts and guyots are volcanic cones that have been built up from the ocean floor by successive submarine volcanic eruptions. There has been much speculation about the flat tops of guyots. One theory indicated that the guyots could have been planed off only if they once stood above the ocean surface. But many of these guyots now lie 6000 feet below the surface. Therefore, they are indeed ancient drowned islands. Core sampling and dredging have revealed sandy deposits and corals on the tops of guyots. Such materials would indicate that shallow-water conditions prevailed until the guyots sank beneath the surface. But another question is presented as to

Fig. 5-8. The Continental Margin and Ocean-Basin Floor adjacent to the U.S. east coast. The continental shelf contains many submarine canyons along the northeast section. The largest of these canyons is the Hudson River canyon that extends southeast from New York Harbor to the edge of the shelf and down the slope. The Muir and Kelvin seamount groups are located north of the Bermuda Pedestal and are the most outstanding features on the abyssal plain. (This is a portion of the Physiographic Diagram of the North Atlantic Ocean published by the Geological Society of America © Copyright by Bruce C. Heezen and Marie Tharp. Reproduced by permission.)

their submergence. Did the sea rise to cover these guyots or did the crust of the earth sink—carrying the mountains below sea level? The second process has been favored by most oceanographers. The mountains became too heavy for the ocean floor and began to sink. Waves gradually planed the cones, and coral organisms grew about the margins of the mountain. But the ocean floor sank faster, and the coral organisms were unable to grow faster than the mountain's descent and eventually died. The planed-off mountain sank below the ocean's surface to either a temporary depth or its present depth. The picture resulting from this theory is that of a vast mountain range in the Pacific about 100 million years ago.

The Mid-Ocean Mountain Range

No scientific evidence concerning a mountain range in the middle of the ocean basins existed prior to 1873. In that year the Mid-Atlantic Ridge was discovered by scientists of the *H.M.S. Challenger.* A 200-pound sounding lead on hemp line was lowered at oceanographic stations located about 160 kilometers (100 miles) apart. The true nature of the Mid-Atlantic Ridge and associated topography began to emerge as the result of detailed echo sounding obtained by the German oceanographic research ship *Meteor* in the period 1925–1927. The Ridge was found to be a rugged mountain range. At about the same time, the discovery of other oceanic mountain areas was coming to the attention of oceanographers. Alexander Agassiz aboard the research vessel *Albatross* found what he thought was a broad plateau off the Mexican coast in the eastern Pacific Ocean. Prior to this, soundings by the *Challenger* had suggested the presence of a topographic rise in the same area. Because of its supposed plateau-like features this area of bottom topography was named the Albatross Plateau. During echo sounding operations in the same Pacific area in 1929 scientists of the research ship *Carnegie* found that the Albatross Plateau was in reality a long southward-trending ridge that is now called the East Pacific Rise.

This rise has an average relief of 2 to 3 kilometers (6500 to

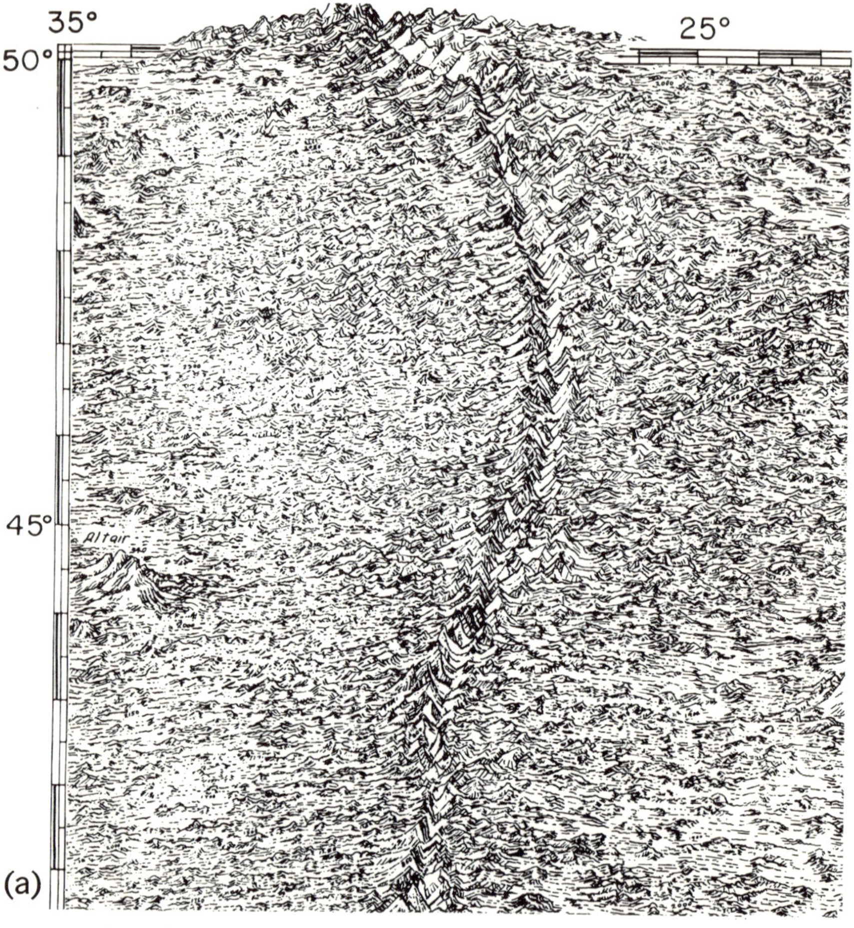

Fig. 5-9. **(a)** Physiographic map of the Mid-Atlantic Ridge showing the rift mountains and the rift valley north of the Azores. **(b)** Physiographic map showing the relationship of the Mid-Atlantic Ridge to the Atlantis, Plato, Cruiser, and Great Meteor seamount groups

9800 feet) and is 2000 to 4000 kilometers (1250 to 2500 miles) wide. It extends from New Zealand a distance of 13,000 kilometers (8000 miles) into the Gulf of California. Oceanographers believe that it originated from a bulge in the earth's mantle. Striking changes in this ridge occur where the east-

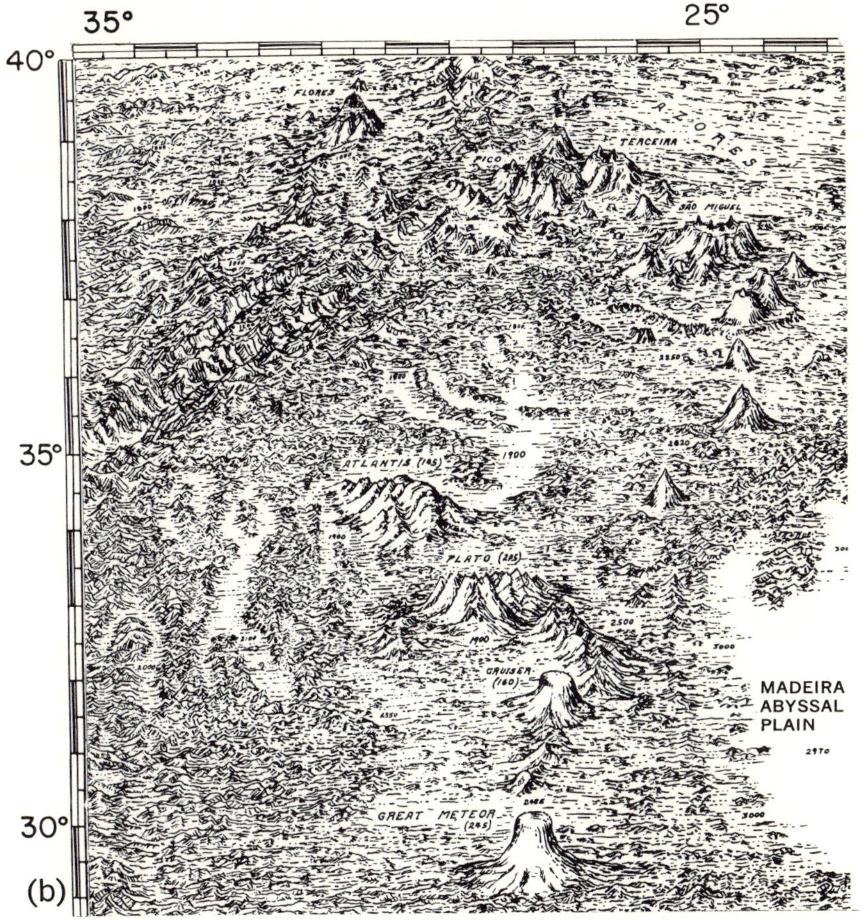

(south of the Azores). (Fig. 5-9 is a portion of the Physiographic Diagram of the North Atlantic Ocean published by the Geological Society of America © Copyright 1957 by Bruce C. Heezen and Marie Tharp. Reproduced by permission.)

west trending fracture zones of the eastern Pacific cross the rise. These zones are great faults that extend thousands of miles west of the California coast and have an average width of 60 miles. Fracture zones in the ocean basins represent breaks in the oceanic crust in which movement of rocks on either side of the break has occurred. The northernmost zone, the Mendocino fracture zone, suggests a possible connection with the San Andreas fault, a land fault. The Murray fracture zone appears to be a continuation of the east-west ranges of southern California. The Clarion fracture zone has not been fully explored. It seems to be connected with the east-west volcanic belt of southern Mexico and possesses volcanic islands along its length. The Clipperton fracture zone also has not been completely explored. East-west fracture zones have recently been discovered in the equatorial Atlantic.

In 1956 Maurice Ewing and Bruce Heezen of Lamont Geological Observatory began plotting the known position of mid-ocean ridges. It was found that most shallow oceanic earthquakes had occurred below the crest of the known ridges. A shallow earthquake is one that occurs within the upper 70 kilometers (43 miles) of the earth's solid surface. Continuing to plot the distribution of shallow oceanic earthquakes, Ewing and Heezen were able to predict that a continuous range of oceanic mountains would be found circling the earth. From the available scientific evidence it is concluded that a continuous mountain range that may be a few hundred miles to a thousand miles wide circles the earth for a distance of 40,000 miles and reaches into every ocean basin. This mountain range alone occupies about one-third of the whole Atlantic Ocean area. In general the crest of the ridge is found less than 2000 meters (6500 feet) below the ocean surface.

Every research cruise helps to confirm the bold conclusion that indeed such a mountain range exists. It was only in March, 1960, that the continuity of the mid-ocean ridge in the western Indian Ocean was confirmed by echo sounding. Prior to this discovery there was speculation that the mid-ocean ridge stopped southeast of Madagascar (Malagasy Republic) and did not join the Carlesberg Ridge (discovered in the 1920's by the Danish research vessel *Dana*) that swings south from the Gulf of Aden and extends southeast into the Indian Ocean. Extension of the mid-ocean mountain ridge is shown in Figure 5-10. It will be noted that no ridge has been found down the middle of the North Pacific Ocean Basin.

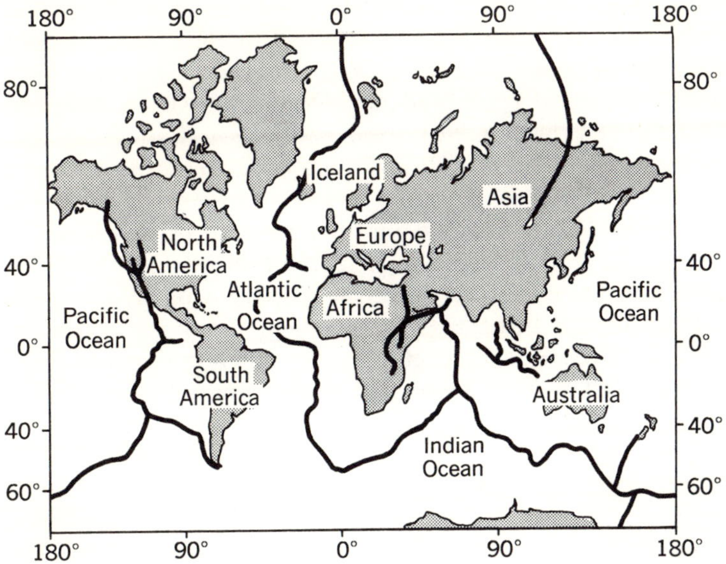

Fig. 5-10. The mid-ocean rift system based on soundings and earthquake records. The entry of rifts into the Red Sea and the Gulf of California suggest new oceanic basins might be forming in those two areas.

The present picture of the mid-ocean mountain range is most impressive. A mid-ocean ridge crosses the Arctic Ocean and joins with the Mid-Atlantic Ridge east of Greenland. The Mid-Atlantic Ridge averages about 1660 meters (5450 feet) above the ocean basin. It begins near Jan Mayen Island north of Iceland to Bouvet Island in the South Atlantic. Some of the mountain peaks extend above sea level and are inhabited by limited plant and animal life. St. Helena, St. Paul's Rocks, Ascension, Azores, and Tristan du Cunha are some of these mountain peaks (see Fig. 5-9b). The ridge is broken in some areas as it extends south toward the Antarctic—such as the east-west Romanche Trench near the equator. The ridge turns between Africa and Antarctic and joins with a ridge east of Madagascar in the Indian Ocean. This is the Carlesberg Ridge which has landward extensions through the Gulf of Aden, the East African Rift Valleys, and toward Pakistan. The main Indian Ocean Ridge continues south of Australia and joins with the East Pacific Rise near Easter Island. As it extends into the Pacific Ocean, the ridge has three landward extensions: from Macquarie Island into Tasmania, New Zealand, and the Ross Sea. From Easter Island, the ridge continues to the Line Islands and then to Kamchatka. At Easter Island, the ridge has two landward extensions: from Easter Island into Southern Chile and into the Gulf of California.

Since the end of World War II, the Mid-Atlantic Ridge has been intensively studied by oceanographers from Woods Hole Oceanographic Institution and Lamont Geological Observatory. By 1953 enough precision echo-sounding data had been accumulated so that Marie Tharp and Bruce Heezen of Lamont were able to draw a physiographic diagram of the North Atlantic Ocean Basin (Figs. 5-8 and 5-9). The diagram showed a startling feature that had not been obvious on the individual profiles. A *rift valley* (also called a *cleft* or *graben*) was found to extend along the crest of the Mid-Atlantic Ridge (Figs. 5-3a, b, c, d; 5-8; and 5-9). The floor of the rift valley averages from 1280 to 3840 meters (4200 to 12,600 feet) below sea level. Its width varies from 20 to 30 miles. The rift valley is bounded on both sides by a line of mountains that rise an average of 1820 meters (6000 feet) above the adjacent rift valley floor. These mountains are called *rift mountains.* The place in the earth's crust or mantle where earthquakes occur is called the *focus* of the earthquake. The place on the earth's surface directly above the focus is called the *epicenter.* When the epicenters of shallow oceanic earthquakes were plotted on the diagram of the Mid-Atlantic Ridge, most of the epicenters coincided with the rift valley. This is more amazing when we realize that the flanks of the Mid-Atlantic Ridge extend 1600 to 1900 kilometers (1000 to 1200 miles) across the Atlantic. Since the rift mountains are a maximum of 48 kilometers (30 miles) apart, the overlap of epicenters and rift zone is striking. This parallelism of epicenter and rift zone has generated much scientific speculation about the nature of the rift zone. From a scenic standpoint the rift valley is twice as wide, twice as deep, and many times more impressive than a section of the Grand Canyon. There are several smaller rift valleys extending parallel to the crest of the East Pacific Rise.

If we want to see continental topography that is an example of the crest of the Mid-Atlantic Ridge, we must observe the terrain of Iceland. This island, which is between Greenland and the European continent, sits astride the Mid-Atlantic Ridge. It has been built by volcanic lava flows similar to those occurring on the mid-ocean mountain crests. There are more active volcanoes on Iceland than on any other land area of comparable size. There have been at least 40 volcanic eruptions on the island since it was settled 1100 years ago. One volcano, Mount Hecla, has erupted 23 times during that period. On November 14, 1963 a submarine eruption began just south of the Westmann Islands, which lie about 32 kilometers (20 miles) south of Iceland (see Fig. 5-11). By May 21, 1964, this new volcano had grown into a 1.6-kilometer (1-mile)

Icelandic Airlines

Fig. 5-11. Surtsey, an active volcano located on the crest of the Mid-Atlantic Ridge south of Iceland. Biologists and geologists studied this volcanic island four months after its eruption and found no form of life. In late October 1964, approximately one year after the eruption, some forms of life (a moth, a mussel, a flowering plant, and seaweed) were found in various parts of the island. These discoveries will enable the ecologist to study the establishment of organisms on volcanic islands.

long island rising 180 meters (600 feet) above sea level. Arising from a volcanic eruption along the crest of the Mid-Atlantic Ridge, this new island has been named Surtsey, from the god of fire and destruction in Norse mythology, Surtr (the Black One).

Extending roughly north-south through Iceland is a great rift valley that corresponds with the rift valley of the Mid-Atlantic Ridge. This valley is called the Central Icelandic Graben (Fig. 5-12). As in the case of other portions of the Mid-Atlantic Rift, the Central Icelandic Graben is becoming wider as time passes.

The discovery of the mid-ocean ridges has increased interesting speculation on the origin and position of the continents. This discovery revived the hypothesis of drifting continents. It was postulated that the present continents were originally joined together in one or two large masses. These masses broke up and slowly floated apart. The present continents possibly are still drifting away from each other. If the drift explanation were correct, dredging should bring up from the mid-ocean

Icelandic Airlines

Fig. 5-12. Almannagja, a 37-meter (120-foot) deep graben in which the first Icelandic Parliament met in 930 A.D. A road now runs along the floor of the graben between the steep walls of the valley. This small rift valley is part of the Central Icelandic Graben, where the mid-ocean ridge comes above the surface.

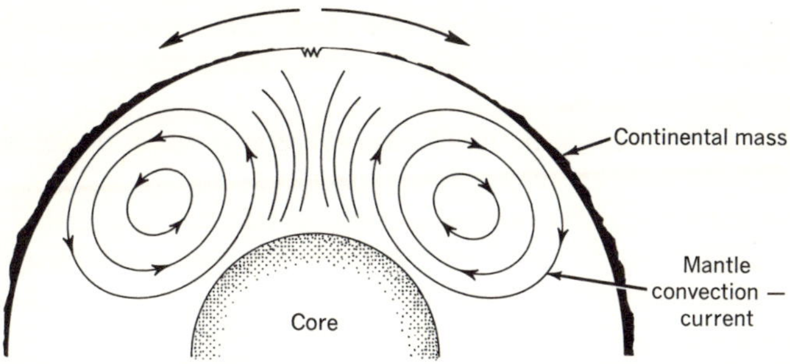

Fig. 5-13. A simplified diagram of the convection-current theory.

ridges the typical continental granitic, acidic-volcanic, and sedimentary rocks. But on the contrary, oceanographers have found serpentine, peridotite, and gabbro rock as well as basaltic lava. If the continents have drifted to their present positions, the mid-ocean ridges would be inactive and a long-abandoned rubble pile. This is not the case since the mid-ocean ridges are very active and are sites of volcanic action. The theory of continental drift had been placed in mothballs until a new scientific tool known as *paleomagnetic analysis* was devised. Paleomagnetists measure the orientation of magnetized particles in rock of the continents. From paleomagnetic evidence, it has been estimated that North America and Europe have drifted apart approximately 2000 miles in the past 300 million years.

Bruce Heezen of Lamont Geological Observatory has revived the concept of an expanding earth in order to explain the apparent continental drift. He suggests that additional crust is being formed in the oceans so that the earth expands and the continents remain the same size. The formation of additional crust is actually occurring in the mid-ocean ridges.

Another theory, the convection-current theory, postulates that the ridges had their origin in convection currents within the earth. Convection currents in the viscous material of the earth's interior rise to the crust under the mid-ocean rift. The movements of the currents would pull the rift apart, and material would be forced through the crack (see Fig. 5-13).

These movements would add material to the underside of the continents and actually compress the continents—resulting in mountain formation.

These theories have led to concrete programs of scientific investigation from which we may eventually be able to develop a theory that fits most and perhaps all of the facts. This increased scientific exploration will add to our present knowledge of submarine topography.

6

Oceanic Sediments and Turbidity Currents

Sediment from the Continents

As we saw in Chapter 2, the rivers that drain water from the continents bring the products of rock weathering to the ocean. Part of the material carried by a stream (the stream's *load*) consists of material dissolved in the water. This is the *invisible load* of the stream. Most of the "salts" in the ocean were carried there as invisible load of streams. The *visible load* consists of fine fragments, such as clay and silt, carried in *suspension*, and coarser fragments that skip or roll along the bed of the stream. These fragments traveling in contact with the bed of the stream are called *bed load*. Some streams are a muddy brown color because of the suspended clay particles, while others are a milky white.

Where the main stream reaches the ocean, much of the load is deposited very close to the shoreline because the ocean water is not moving fast enough to carry it way from the mouth of the river. The deltas at the mouths of major rivers are evidence of this deposition. The coarser particles of the bed load are usually deposited close to the river mouth, whereas the suspended load may be carried seaward for many miles before it settles to the bottom. In this way the delta continually builds outward.

Waves, wave-induced currents, and tide-induced currents carry some of the deposited material away and build beaches elsewhere along the shoreline. These beaches are usually composed primarily of white quartz sand; the brownish, clay fragments, because of their small size, are carried further out to sea.

When oceanographers first calculated the amount of sediment brought to the shoreline by streams and then compared this quantity with the amount of sediment found in beach, delta, and continental-shelf deposits, they were surprised to find that much of the stream sediment had disappeared. Could waves have carried it beyond the continental shelf and into the deep ocean? Generally speaking, the answer to that question must be "No." What about ocean currents, especially currents running along and across the continental shelf? We have already mentioned that such bottom currents exist and could transport sediment to the continental slope. But even this explanation does not account for all the missing sediment. It was not until 1952 that Bruce Heezen and Maurice Ewing of Lamont Geological Observatory were able to combine many lines of evidence in an explanation of some deep-ocean sediments and their relation to continental-shelf sediments.

Submarine telegraph cables and, in more recent years, submarine telephone cables, play an important role in the theory of oceanic sedimentation offered by Heezen and Ewing. The first transatlantic telegraph cables were laid in the 1850's with the expectation that they would remain undisturbed and trouble-free for many years of profitable use. It was thought that the oceanic abyss was a place of silence and calm. At first it appeared that the cables would be slowly covered by dust-like particles of atmospheric debris and planktonic remains that settled out from the ocean surface above. Rapid and frequent cable breaks soon shattered that illusion. What had caused the cable breaks? Many cables were severed by the action of marine organisms. Heezen reports that sperm whales were definitely responsible for 14 of the breaks reported by cable companies. Dead whales with cables wrapped around their jaws and flukes have been recovered as confirmation of their role in the cable failure. It has been of interest to biologists to learn from these breaks that sperm whales became entangled in the cables while plowing through bottom sediment at depths as great as 1134 meters (3720 feet). This is indirect evidence of the maximum depth reached by sperm whales. But the many hundreds of cable failures cannot all be blamed on whales.

Some cable breaks occurred very close to earthquake epicenters (the point of maximum earthquake-intensity on the earth's surface). Rapid motion of bottom sediment and rock, resulting from the earthquake shock, subjected the cables to a greater strain than they were built to resist. But again we find that neither sperm whales nor nearby earthquake epicenters could be responsible for all cable breakage. Nonetheless, one particular earthquake set off a chain of events that led to a more complete understanding of cable breaks and processes of erosion and the deposition of oceanic sediment. This earthquake was the now famous Grand Banks earthquake that occurred on the continental slope just south of Newfoundland at 8:32 PM on November 18, 1929. Eight cables near the epicenter, high up on the continental slope, were broken instantly by the earthquake. Five other cables lying further down the slope (actually on the continental rise and Sohm abyssal plain) broke later in a seaward succession. The first of the five broke 59 minutes after the quake. The last cable, lying about 650 kilometers (400 miles) seaward of the epicenter, broke 13 hours and 17 minutes after the earthquake. Heezen and Ewing reasoned that a mass of sediment lying near the epicenter had been loosened by the earthquake and had flowed rapidly down the continental slope, cutting the cables as it traveled. Such a mass of dense, sediment-laden water flowing down-slope in contact with the bottom is called a *turbidity current.* Like an avalanche of snow on land, these currents may reach great speeds. From the records of the Grand Banks cable breaks it has been calculated that the turbidity current was traveling at 102 kilometers per hour (55 knots) when it cut the first cable on the upper part of the continental rise. The current had slowed to 22 kilometers per hour (12 knots) by the time it cut the last cable lying on the gentle seaward slope of the Sohm abyssal plain. The down-slope succession of cable breaks resulting from the Grand Banks earthquake and a similar sequence of down-slope cable breaks associated with the Orléansville, Algeria, earthquake in 1954, gave convincing evidence that turbidity currents can be a powerful eroding force.

Additional proof of cable breakage by turbidity currents was obtained when sediment cores from the abyssal plain south of the Grand Banks revealed *graded beds* of sand and silt in the area where the supposed turbidity current had slowed to a halt. That is, the coarser sediment particles were concentrated at the bottom of the layer, whereas the finer particles were concentrated near the top of the layer. It had been suspected that a great deal of sediment was involved in the turbidity current, because the cable companies had grappled unsuccessfully for the severed ends of many cables. They had correctly assumed that, as in similar cases, the cable ends were buried in sediment. But they had not known just how the sediment was deposited on top of the cables. In the light of the evidence, it was reasoned that the abyssal plains were probably created primarily by deposition of sediment carried by turbidity currents. This was indeed a striking clarification of the relationship between oceanic topography and sedimentation.

In addition to cable breaks caused by turbidity currents, is there additional evidence that these submarine flows can erode the ocean floor in the way rivers erode the continents? Is there additional evidence that abyssal plains are primarily deposits of sediment carried down the continental slope and continental rise by turbidity currents? Plant remains and shells of shallow water organisms found in sediment cores taken from abyssal plains are explained by turbidity-current deposition. But more impressive evidence of the cutting power and frequency of turbidity currents is shown by the many *submarine canyons* that descend from the shelf break to the base of the continental slope. Natural levees, like the levee of the Mississippi River that is built along the river by sediment carried during floods, run along both sides of the canyons. At the base of the canyon is a delta-like deposit of sediment. Can there be any doubt that the canyons were cut, and that the levees and deltas were built so by the action of turbidity currents? At first it was argued that submarine canyons had been cut by streams that ran down the dry continental slope when there was much less water in the oceans. However, this argument is no longer widely accepted, because oceanographers now doubt that the sea level was ever low enough in the recent past to expose the continental slope to the extent required by the theory. Submarine canyons in the continental slope range up to about 16 kilometers (10 miles) in width and up to about 1800 meters (6000 feet) in depth. Such rivals of the Grand Canyon add a rugged beauty to the topography of the continental margin. The submarine valley of the Hudson River leading almost to the shelf break and the Hudson Submarine Canyon lying seaward of the shelf break, can be seen in the physiographic diagram of the eastern United States presented in the chapter on ocean basin topography. In recent years, the discovery of submarine canyons out in the floor of the deep ocean basins have shown that not all submarine canyons are associated with the mouth of a major river or to earthquakes near the continental margin.

Other Sources of Oceanic Sediment

Sediment-filled basins running along the continental shelf off New England have recently been explored by seismic-refraction methods. In portions of these basins the sediment reaches a thickness of over 6100 meters (20,000 feet). Rivers discharging their load onto the continental shelf are the obvious source of such sediment. Even more recent geophysical investigations in the deep oceans have revealed that the thickest sediment accumulation in any of the abyssal plains is to be found in the Argentine Basin off the east coast of Argentina. The Argentine abyssal plain, lying on the western side of the Argentine Basin, was found to have a sediment thickness of at least 2500 meters (8200 feet). Sediment delivered to the continental shelf by the Rio de la Plata has been released to the abyssal plain by turbidity currents occurring over long periods of time. But what of the sediment found on the mid-ocean ridges and in other areas that lie beyond the reach of turbidity currents?

Oceanic sediments derived from weathering of continental rocks are called *terrigenous sediments.* Abyssal-plain sediments are mostly of terrigenous origin. Sediments that have been precipitated from ocean water or that have been built by the accumulation of organic remains such as the *tests* (i.e., shells) of planktonic marine animals are referred to as *pelagic sediments*. The three most important types of pelagic sediment are the *calcareous oozes, siliceous oozes,* and *red clay*. Calcareous oozes including Globigerina and Pteropod oozes are the most widely distributed pelagic sediments. Globigerina ooze is derived primarily from the calcareous shells of closely related genera of planktonic foraminifera (see Chapter 7). Pteropod oozes represent the calcareous shells of marine mollusks. Electron-microscope studies have shown that the minute calcareous skeletal plates (*coccoliths*) belonging to the pelagic marine flagellate "algae" Coccolithophorida are an important component of the calcareous oozes.

The siliceous coverings of diatoms and radiolaria (Chapter 7), account for most of the volume of siliceous oozes. Diatoms are plant organisms whereas the radiolaria are marine animals that secrete a silica (actually opal, a compound containing silicon dioxide) test. Red clays are considered to be the insoluble remains of organic material (such as the calcareous tests of foraminifera) that has been partially dissolved by ocean water. Fine dust carried from land by ocean currents and wind also accounts for part of the volume of red clay. Meteoric dust settling to the ocean floor adds to the mass of red clay.

In contrast to the rapid deposition of terrigenous sediment by turbidity currents, it has been estimated that, in the deep ocean, calcareous oozes are presently accumulating at an average maximum rate of about one centimeter (about four-tenths of an inch) every thousand years. Red clay is accumulating at about half that rate. Near the continental margins where organic productivity is higher, the general rate of accumulation of pelagic sediment may increase to about 60 centimeters (24 inches) every thousand years. This slow buildup of pelagic sediment helps to explain why the sediment cover of the ocean floor is so thin in many places. Nevertheless, it should be remembered that the amount of time represented by even relatively thin sediment layers assumes astronomical proportions. For example, a very long core of 18 meters (59 feet) brought up from an area of ocean bottom where sediment has been accumulating at a rate of one centimeter per thousand years represents a period of 1,800,000 years. But this is only a small fraction of the earth's age. The deep drilling associated with the Mohole project may allow oceanographers to obtain sediment samples from an earlier period of the earth's history. Many fundamental questions about the age of ocean basins and the beginning of life on earth may be answered by penetrating deeper into the oceanic sediment layer.

Bottom-sampling devices

In the early years of oceanography there were no devices suitable for obtaining large amounts of bottom sediment and rocks from deep ocean basins. A gum was sometimes affixed to the bottom of the sounding lead. A small amount of bottom sediment would stick to it and be returned to the surface. But such primitive methods told the oceanographer nothing about the sediment below the thin surface layer, nor could he even be sure that the sample at the bottom of the sounding lead was representative of sediment in that area. Picture closing your eyes, sticking your arm into a refrigerator, and pulling out the first object you touch. To what extent would it be representative of the contents of the refrigerator? Probably the object would be a foodstuff or would contain a foodstuff. Obviously, classifying the contents of refrigerators in this manner and classifying bottom sediments in the ocean by lead-line sampling both leave much to be desired.

The oceanographer seeks to know whether the ocean bottom consists of outcropping rock or whether it is covered with sediment. He would like to obtain a sample of the outcropping rock for petrologic analysis. A sample of bottom sediment would be examined for the presence of fossil remains. The nature of the sediment, whether clay, silt, sand, gravel, or some combination of these is also of interest. A common device for obtaining large fragments of rocks that might be exposed on the ocean bottom is the *geological dredge sampler* (see Fig. 6-1). When towed behind the ship, the dredge sampler cuts a swath in the bottom sediments, which are strained through the coarse chain mesh hanging below a boxlike opening that is actually the front of the sampler. This method of sampling is called "trawling." Sedimentary rocks are found outcropping from the continental slope and from the walls of marginal trenches. A portion of the sedimentary rock is broken off by the sampler and retained in the chain-mesh basket. Rocks of the mid-ocean mountains have also been recovered from dredges of this type. Keeping the front of the dredge in contact with the bottom requires a great deal of experience. Very often the dredge snags an unyielding rock mass and

Official U.S. Coast and Geodetic Survey Photograph

Fig. 6-1. Geological dredge samplers like the one shown above are used to bring up chunks of ocean-bottom sediment and rock for examination.

the towing cable snaps or adroit maneuvering of the ship is required to free the dredge. In spite of all these difficulties, many valuable samples have been obtained. An attempt to trawl for dredge samples from the Mariana Trench in 1957 failed on three successive tries. This was especially disturbing as it took more than 12 hours to lower and recover the dredge from the bottom of the trench. Dredge-hauls from the area of the Mid-Atlantic Ridge have recently been used to determine that volcanic rocks from which the ridge was formed are basalt of a different composition from that found on volcanic islands, such as the Azores, that rise above the ridge.

A large volume of sediment and relatively small masses of rock from the surface layer of ocean-bottom material can be obtained by use of a variety of devices known as "grab" samplers. When lowered to the bottom and then triggered, the jaws of the sampler close rapidly and scoop up a sample of bottom material. This material is automatically sealed in the sampler for the return trip to the surface. The type of grab sampler shown in Figure 6-2, the "Campbell Grab Sampler," consists of two scoops that snap together to form a semicircular container for the bottom sample. The one pictured in the figure weighs about 318 kilograms (700 pounds) and can pick up one cubic meter (35 cubic feet) of sediment. A camera is mounted inside one of the jaws and a high-intensity strobe light is mounted inside the other jaw. A photograph of the bottom material to be sampled is taken just before the sampler strikes the bottom. This grab-camera sampler has solved one problem that long plagued oceanographers. Photographs of the bottom were not very useful for identifying marine animals shown in the photograph. Ripples on the bottom could be seen on photographs but it was difficult to identify the material constituting the ripples. Grab samples by themselves gave the oceanographer no real idea of what the bottom surface sampled looked like. The obvious solution was to place a camera and light directly in line with the jaws of the grab. This solution has proved difficult to engineers; the first 31 trials were unsuccessful and performance-reliability is still only about 25 percent. Even so, the Campbell Grab has already

Official U.S. Geological Survey Photograph

Fig. 6-2. The Campbell Grab Sampler is used to scoop up large volumes of ocean-bottom sediment. A camera in one jaw of the sampler and a light in the other jaw are used to photograph the ocean bottom before the sample is taken.

become a standard piece of oceanographic "hardware."

Grab-samples are usually too large to be completely saved and stored. Marine geologists and biologists usually take a small sample of the material for storing in a reference museum. The remainder is worked over so that the oceanographer can remove the material in which he is interested. One means of removing the fine sand, silt, and clay fractions of the sample is provided by the shaker table pictured in Figure 6-3. The sample is shoveled onto a horizontal wire screen of appropriate mesh size. The wire screen vibrates and water is sprayed on the sediment at the same time. The fine material is washed through the screen and the coarse material is retained for examination.

Layering in bottom sediments is of the greatest interest to marine biologists, chemists, and geologists alike. Differences in the size of sediment particles, chemical composition and fossil organisms in each layer often provide important information about the physical conditions at the time the sediment was deposited. The *coring tube* is one of the most successful devices used to obtain an undisturbed sample of the ocean bottom sediment layers. Almost all of the coring tube devices in common use employ the "free-fall" principle. A heavy weight attached to the top of the coring tube is allowed to drop when the device has been lowered almost to the bottom. The force of the falling weight pushes the coring tube into the bottom. With luck, the coring tube will penetrate vertically into the sediment. When returned to the ship the sample can either be stored in the coring tube or extruded for later examination (see Fig. 6-4).

Official U.S. Geological Survey Photograph

Fig. 6-3. A shaker table for sieving grab samples.

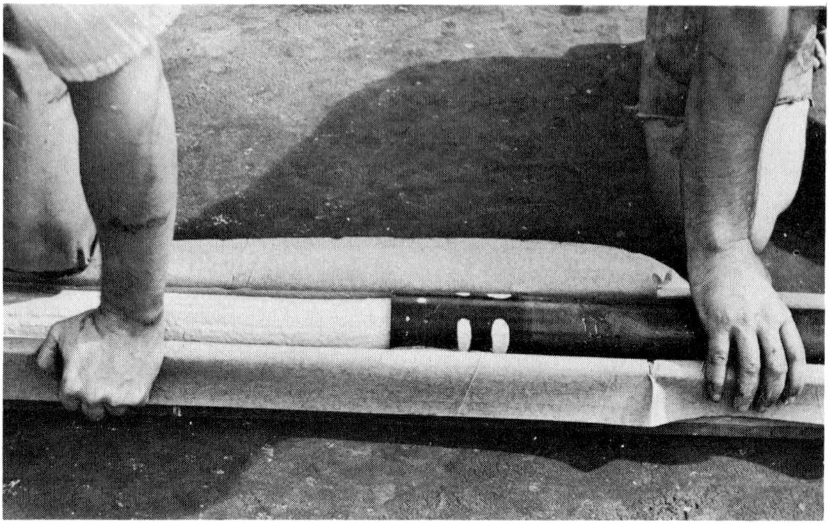

Lamont Geological Observatory

Fig. 6-4. Scientists aboard the *Vema* extrude a sample of bottom sediment from a cording tube.

The "Phleger Corer" is used to obtain cores of bottom sediment up to about one meter (39 inches) in length. This device consists of a steel coring tube attached below a main body weight. A tailfin assembly above the main body weight stabilizes the corer during descent. Phleger corers employ the free-fall principle mentioned above. A plastic liner inside the tube results in a sample that can be removed and stored in its own plastic sleeve without the necessity of extruding it into another wrapper.

Another corer employing the free-fall principle is the "Ball Breaker" corer shown in Figure 6-5. A Phleger corer is attached below a cylindrical chamber that contains a glass sphere. When the corer is triggered the glass ball is broken and the sound is picked up by a hydrophone. In this way the topographical record from the echo sounder can be matched against the sediment.

The Ewing piston corer shown in Figure 6-6 is different from the Phleger corer in that a piston in the tube helps to pull the core into the core pipe. Cores up to 18 meters (59 feet) in length have been recovered from all ocean basins with this coring device. As there is no plastic liner in the coring tube the cores must be extruded, cut into appropriate lengths, and wrapped for storage. This procedure benefits the oceanographer during long cruises because the core can be sampled before wrapping and the samples can be studied while the ship is under way.

Coring devices are constantly being improved so that vertical cores, completely undisturbed cores, deeper cores, and larger cores can be obtained. More efficient "core catchers" are being designed to prevent the core from slipping out of the tube on the way to the surface. On deep sea cruises it is diffi-

Official U.S. Coast and Geodetic Survey Photograph

Fig. 6-5. Ball Breaker coring device being lowered over the side.

Lamont Geological Observatory

Fig. 6-6. Ewing Piston Corer is lowered over the side of the *Vema.*

cult to obtain even one moderately long core in one day of work. This is because the ship must remain motionless during the many hours required to lower and raise the coring device. Many cores are lost on the way up to the surface and the corer frequently strikes rock instead of sediment. Most oceanographic institutes have collected only a few thousand long cores—perhaps in the entire world there are only 10,000 or so long cores in storage.

Photographing the Ocean Bottom

Underwater photography is a most useful adjunct to bottom sampling for biological and geological specimens. A photograph can tell the oceanographer whether it was the turbulence of ocean waves or an underwater current that has rippled the bottom sediment. Turbulence of ocean waves produced the rippled bottom pictured in Figure 6-7.

A large proportion of photographs of the ocean floor reveal the presence of asymmetrical ripples in the bottom sediment. Like the sand ripples produced when water runs from tidal pools on the beach, these asymmetrical or "current" ripples have a steep face in the direction in which the water flows. Because of this similarity, oceanographers believe that swift-moving water currents are to be found along the ocean bottom. These are not sediment-laden turbidity currents, but simply swiftly moving currents of "clear" water. Current-ripple marks and evidence of scour by water currents have been found on every seamount photographed to date. Rippled sediments have been photographed even on the floor of the rift valley in the Mid-Atlantic Ridge. Shells and gravel lie in the ripple troughs, whereas the fine sediment is found on the crests of

Official U.S. Geological Survey Photograph

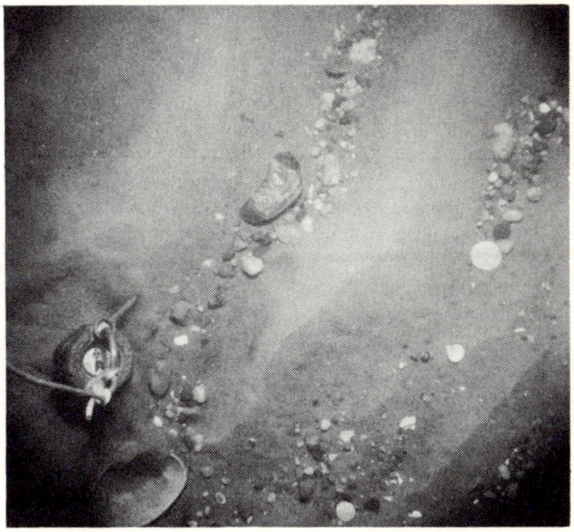

Fig. 6-7. Rippled bottom sediment at a depth of 20 meters (65 feet) a few miles south of Long Island, New York. The camera trip weight at lower left has a compass attached, permitting the orientation of the ripple crest to be determined.

the ripples. These ripples are fairly symmetrical (i.e., the sides of the ripple slope downward from the crest at the same angle). This indicates that there was no strongly directional current sweeping over the bottom.

The first serious attempt to photograph the ocean bottom with an automatic camera was made in 1893. Even though it was made in the clear waters of the Mediterranean Sea, the amount of illumination available was too little for the relatively insensitive film of that time. It was 47 years later that a suitable automatic underwater camera and light were designed by scientists at Woods Hole Oceanographic Institution. Since 1940 this camera unit received extensive use in photographing the bottom. Prior to that time, the only bottom photographs where those taken in shallow water by divers.

The first automatic underwater cameras could take only a single photograph at each lowering to the ocean bottom. Constant improvement in underwater lights and camera mechanisms has resulted in versatile equipment like that shown in Figure 6-8. The two cylinders at the top of the housing hold cameras. The cylinder and disc in the center is a sonar-pinger unit that allows the operator to adjust the length of line so that the cameras remain at the proper distance above the bottom. The four cylinders at the base of the unit contain underwater lights and batteries. In contrast to the single photograph per lowering obtained by earlier automatic cameras, these cameras can take up to 500 pictures per lowering.

Bottom photographs also record the tracks and trails of bottom-dwelling marine organisms. Occasionally bottom-living plants and animals show up in the photographs. The presence of life in the Mariana Trench was already known, before the descent of the bathyscaph *Trieste,* from a photograph of a brittle star (similar to a star fish) at the bottom of the trench. Many taxonomic classifications of bottom-dwelling marine animals have to be made from the photographs because many of these creatures have eluded the nets sent down to catch them. If an electrically operated biological sampler with a television camera just back of the mouth of the sampler can be designed,

Woods Hole Oceanographic Institution

Fig. 6-8. Underwater camera unit being lowered over the side of the *Atlantis II.*

then the marine biologist may be able to increase the selectivity of biological sampling programs.

Manganese Nodules

Bottom photographs have also brought to light dense concentrations of manganese nodules on the bottom of most ocean basins. (See Fig. 6-9). Such nodules are a potential source of metal ore. These nodules are concretions (for *conc*entric acc*retions*) having a high percentage of iron and manganese in the form of oxides and hydroxides. Manganese nodules are thought to form in much the same way as pearls form in oysters. That is, the manganese and iron dissolved in sea water

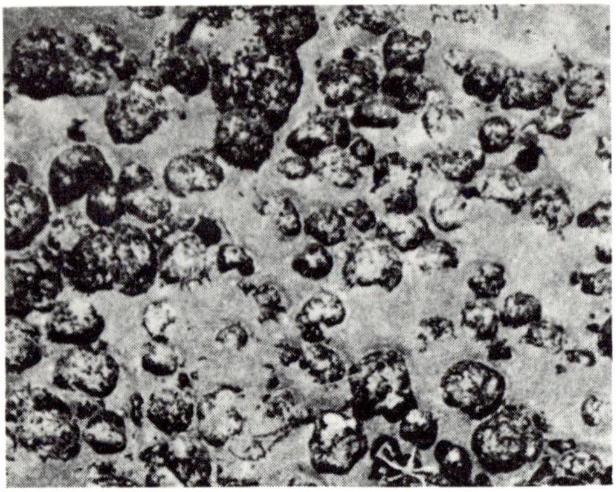

Official U.S. Geological Survey Photograph

Fig. 6-9. A dense concentration of manganese nodules on the floor of the Blake Plateau, at a depth of 823 meters (2700 feet).

will, under favorable chemical conditions, precipitate on a grain of sand or a shell fragment or some other convenient solid material. The rate of deposition on the nodules is unbelievably small. It has been estimated that it takes a thousand years to build a layer of one millimeter thickness. In oceanic areas having a smaller concentration of dissolved iron and manganese, it may take 100,000 years to precipitate one millimeter of iron and manganese oxides and hydroxides. On the average, 100 grams of nodule material would yield approximately 24.2 grams of manganese and 14 grams of iron. While these are important ore materials in themselves, the nodules also contain small amounts (tenths or hundredths of grams in our example) of even more valuable ore material such as copper, nickel, cobalt, zirconium, and molybdenum, to mention only a few.

Even though the trace-element content of manganese nodules may seem small, the concentration of manganese nodules in some oceanic areas is so high that profitable recovery of trace elements would be assured. According to present estimates, 20 to 50 percent of the floor in the southwestern Pacific Ocean is covered with these nodules. There may be up to 70 million kilograms of nodules per square kilometer (equivalent to 200,000 tons per square mile) of ocean floor in some areas. Metals recovered from these nodules might be worth $100 per ton to the nation or company able to mine them.

7

Marine Biology

Life in the Oceans

Since the day man was able to float in a body of water, his descriptions of edible organisms and sea monsters have kindled the naturalist's imagination. As coastal civilizations developed, these edible organisms became an essential item in the human diet. The naturalist thus had an opportunity to study important features of coastal organisms—their life histories, structure, and importance in the food chain. What remained a mystery was the open sea—its depths, its currents, and its inhabitants. This mystery was enhanced not only by the tales of seafaring men but also by the erroneous investigations of many well-known naturalists.

During the 1840's Edward Forbes, a British oceanographer, contributed many new ideas for dredging techniques and enhanced studies on the topography of the sea floor. He was influential in training of future oceanographers and design of future ocean studies. However, his fame rests chiefly on his theory of life in the oceans. Forbes had dredged the waters of Great Britain and the Mediterranean and found that the abundance of plant and animal life decreased at greater depths. He surmised that no life could resist the great pressures at these depths. From these observations, he concluded no marine life existed below 550 meters (1800 feet) and termed the deep ocean *azoic* (without animal life). This theory was widely accepted for a great number of years. However, Forbes' theory was shattered by the laying of submarine cables during the 1850's. One particular cable was hauled up for repair from a depth of 2000 meters (6500 feet), and was found to be encrusted with 15 varieties of living organisms. Deep-sea coral, mollusks, worms, and other sessile organisms were adhering to the cable.

Prior to this event, minor and scattered accounts of discovering life at great depth had been recorded. During the years 1839–1843, the Antarctic expeditions under the direction of James Ross dredged animals from depths of 900 meters (2700 feet). In the Baffin Bay expedition of 1818 under John Ross, an animal was hauled from a depth of 1600 meters (4800 feet). During the 1850's, George and Michael Sars dredged animals from depths of 900 to 1100 meters (2950 to 3600 feet). These scattered accounts were overlooked by many contemporary scientists who claimed that such organisms were caught in the surface waters as lines were hauled aboard.

After the amazing discovery of life encrusted on cables and the results of the *Challenger* expedition, there could be no doubt that life existed in the "azoic" zone of the ocean.

Marine Environments

Walking from the beach into the water and surveying the floor beneath the water surface, we first would encounter strong tides and waves that wash over the beach or pound the rocky coast. At this point, the marine environment is termed *intertidal* (see Fig. 7-1). This zone is part of the *eulittoral* region and covers the sea bottom from low water to high water of the tide. Many species of marine organisms are found only in the tidal zone and may be limited vertically. Many mobile animals, for example crustaceans and fishes, move regularly into the intertidal zone to feed during high tide. In the exposed areas of the tidal zone where there is either direct ocean waves or periods of dessication, the marine organisms are limited and possess specific resistant structures. Barnacles, crabs, limpets, and snails are a few of the marine species found in this zone. Plants (*Laminaria, Fucus*) that possess "holdfast structures" to withstand the force of the waves are also found in this zone.

We then pass into the general *littoral system* which is a part of the *benthic division.* The benthic division encompasses the

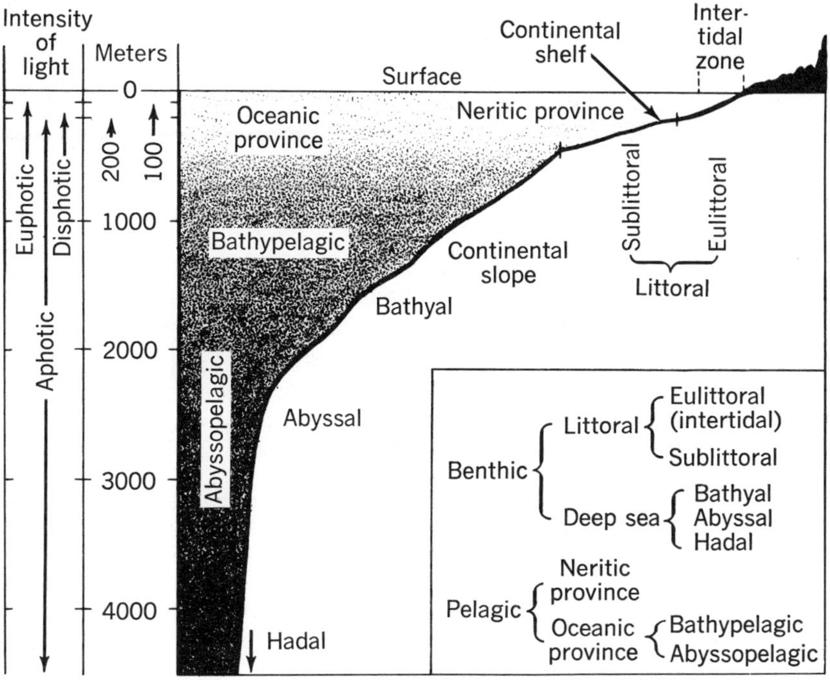

Fig. 7-1. A simplified drawing of marine environments and their relationship to the light zones.

sea bottom from the tidal zone to the greatest depths. The littoral system is subdivided into the *eulittoral,* with the *intertidal,* and the *sublittoral* zones. The eulittoral zone extends from the high tide level to a depth of about 40 to 60 meters (130 to 195 feet). Its lower border is the lowest limit in which holdfast plants can grow. The sublittoral zone stretches from the 60-meter level to a depth of about 200 meters, or the edge of the continental shelf. At this level, we reach a very important feature of oceanic life. The abundance of animal life and more specifically plant life depends on the intensity of light in sea water. This intensity decreases with depth and, of course, varies with the latitudes and the seasons (Fig. 7-1). The sea has been divided into three zones with respect to the amount of light present. The *euphotic zone* is abundantly supplied with light sufficient for plant photosynthesis. This zone extends from the surface to about 80 meters (260 feet). The eulittoral zone lies within this light zone. The *disphotic zone* is dimly lighted and extends from the 80-meter level to about 200 meters, or the edge of the continental shelf. Plant production is not effective in this layer. This layer covers the sublittoral zone. This 200-meter level is a demarcation for not only the life in the ocean but also for the remainder of the benthic division.

The benthic division below the continental shelf is termed the *deep-sea system* which covers from 200 meters to the sea floor. This is a lightless region (*aphotic zone*) in which no plants are produced and animal life is limited. The deep-sea system is divided into an upper (*bathyal*), a middle (*abyssal*), and a lower (*hadal*) zone. The bathyal zone extends from the continental shelf to a depth of 1100 to 2000 meters (3600 to 6500 feet). The bottom in this zone may be rocky or muddy. The organisms are of varying colors and migrate to the rich upper levels at night. The abyssal zone comprises the deep-sea system below 2000 meters to about 6000 meters (6500 to 19,700 feet). From 6000 meters to the deepest trenches, the deep-sea system is termed hadal. These are regions of uniform conditions. There are no seasons, and no plants are produced. To touch lightly upon the features of the animals at the bottom, we would find them carnivorous, feeding on organic debris that must have originated from the organisms of the upper layers.

The ocean water that covers all these regions is called the *pelagic division*. It is divided into the *neritic province* (inshore) and the *oceanic province* (open-sea). The oceanic province contains all the three light zones previously mentioned. The boundary between the lighted and lightless zone is arbitrarily set at 200 meters, since this would correspond with the depth for the edge of the continental shelf. The outstanding features of this province are the spatial expanses and the great ranges of depth. The *neritic* waters cover the eulittoral and sublittoral regions. These are the waters from the edge of the continental shelf to the coastline. It is also the water region that has most of the light intensity and the most sea life.

Plankton Organisms

The abundance or absence of free-floating microorganisms (*plankton*) can be determined by the coloration of the sea. What is thought of as beautiful to humans, the deep blue color of certain regions of the ocean, would be analogous to the desert environment on land. Suspended life would be at a minimum, and only specific endemic organisms would be found. An example of this oceanic region is the Sargasso Sea, named from the floating seaweed *Sargassum.* Copepods, snails, carbs, and worms live among the weeds and would perish outside this environment. But the ocean surface is not the only inhabitable level. If we were to descend below the surface to about 100–200 fathoms (600–1200 feet), we would encounter along the continental shelf and in the open sea free-swimming organisms (*nekton*). This group comprises the fishes and larger marine organisms that feed upon the plankton and other small organisms. The third marine form (*benthos*) can be found on or in the bottom sediment.

Plankton is taken from the Greek word meaning *that which is made to wander or drift.* Planktonic life is divided into two groups, the *phytoplankton* (plant forms) and the *zooplankton* (animal forms).

Phytoplankton. Phytoplankton (see Fig. 7-2) depend upon sunlight for growth and survival. Through the process of photosynthesis, these tiny organisms produce food and energy for the zooplankton and the larger marine organisms. Phytoplankton are usually found a few meters below the surface during periods of most intense sunlight and are very close to the surface at night. Because the intensity of the solar radiation decreases greatly with each small increase of depth, this slight depth-adjustment is believed to be a mechanism by which the phytoplankton avoid the direct intensity of the sun. Just as humans use suntan lotion to avoid overexposure, these tiny floating plants are protected by a water layer between themselves and the sun's rays.

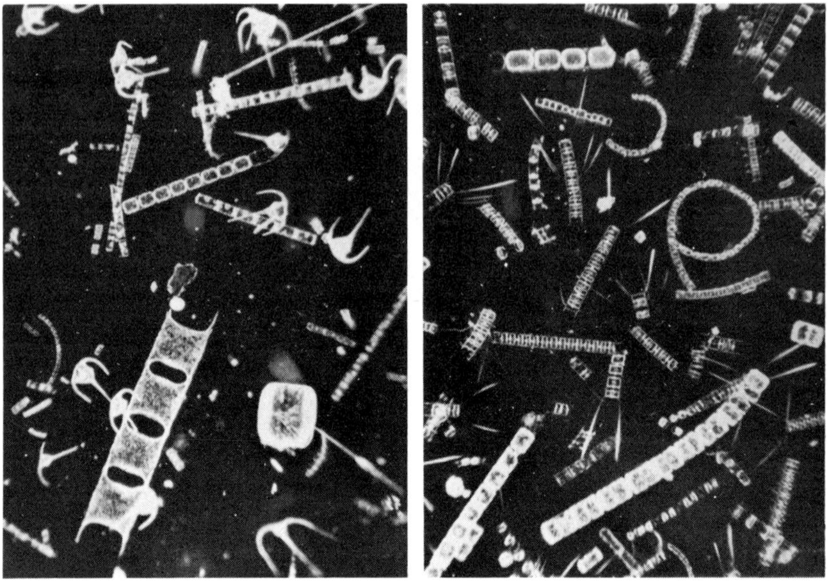

Woods Hole Oceanographic Institution

Fig. 7-2. These organisms are predominantly diatoms which are vital in the food chain of oceanic life.

Whereas knowledge of depth changes of plankton has come primarily from trawling with fine-mesh nets, evidence for a daily rhythm of depth change among some larger marine organisms has come as a result of sonar (echo sounder) surveys of ocean bottom topography. Reflecting horizons were found on the sounder records at depths that are much shallower than the bottom. During the daytime, in most oceans, reflections were received from depths of 213 to 732 meters (700 to 2400 feet). Because of the relatively great depth from which the sonar echoes were reflected and scattered, these reflecting horizons were called *deep scattering layers* (*DSLs*). An intensive investigation of DSLs by Robert S. Dietz, U.S. Coast and Geodetic Survey, revealed that the scattering layers rise to the surface on dark nights and almost to the surface on moonlit nights. Echo traces made in the daytime reveal the shallowest layer at about 215 to 310 meters (700 to 1000 feet). The second layer is at about 450 meters (1475 feet) and the deepest at 560 to 630 meters (1700 to 1900 feet)—and as deep as 800 meters (2600 feet). These three layers merge into a single band as much as 130 meters (425 feet) thick when they move toward the surface.

It has been established that deep scattering layers mostly represent a dense concentration (perhaps as many as 20 per cubic centimeter) of marine organisms that are 0.5–1 centimeter (up to ½ inch) in length. Lantern fish, sergestid worms, and marine crustaceans (euphausids) are thought to be the organisms of greatest abundance in the deep scattering layers. They feed primarily on phytoplankton during their nocturnal ascent to the ocean's surface. Larger fish may prey on the organisms of the DSLs and may also contribute to the density of these layers. Woods Hole oceanographers have invented a camera that sends out a "ping" and takes a picture automatically upon receiving a return echo. As daylight approaches, the lantern fish descend first, then the euphausids, and finally the sergestids. The three layers of the DSLs never cross. Siphonophores also have been known to cause sound scattering.

The most abundant organisms of the phytoplankton are the yellow-green algae, the *diatoms,* which number about 15,000 to 20,000 species. They range from twenty-five ten-thousandths to about one-tenth of an inch in length. These unicellular organisms differ from all other algae in having a cell wall that forms a siliceous external skeleton from minerals in the ocean. Their shape resembles a glass pillbox in which one half of the shell overlaps the other half. These shells are an important component of siliceous sediments and have formed great fossil deposits known as *diatomaceous earth.* About 280,000 tons of diatomaceous earth are produced in the United States each year. It is used for filters, as a base for toothpaste and powder, for fine car polishes, and for insulation.

In order to maintain their existence in sea currents and to avoid sinking to the depths, the diatoms have unique adaptive structures. Several types of devices are the bladder structure, needle or hair appendages, and ribbon and branched formations. The presence of oil drops in the cells also maintains flotation of the phytoplankton because oil is less dense than water.

It is useless to estimate the number of diatoms in the sea, especially with 15,000 to 20,000 species recorded. When the sea environment is favorable and production at its peak, there are diatoms in every drop of sea water in temperate zones. One diatom may have 100 million descendants in a month. The water beneath one square meter in the Gulf of Maine contained 7 to 8 billion diatoms during peak production; 120,000 diatoms have been consumed in one day by a copepod (a crustacean), and a single copepod may consume many millions before its death. Estimates indicate that there are about 4 million tons of copepods existing at any one time in the 36,000 square miles of the Gulf of Maine. One can only try to imagine the number of diatoms present.

With the exception of planktonic bacteria, the flagellates comprise the remainder of the phytoplankton. The dinoflagellates, a group in the flagellates, are second to the diatoms in the economy of the sea. These organisms are more common in warm waters than in cold waters.

These unicellular forms have skeletons made of little plates of cellulose that may carry spines and wings (see Fig. 7-3). Dinoflagellates are propelled through the water by the vigorous waving motion of two whip-like structures called *flagella.* Some phytoplankton, especially dinoflagellates, can be very destructive and may possess strange biological properties. The spring and summer oceans of yellow, green, brown, and red point to the presence of significant and specific algal plants of which some are extremely poisonous. In the Gulf of Mexico, there is usually a late summer bloom of a red-pigmented dinoflagellate, *Gymnodinium brevis,* which comprises the disastrous "red tide." A tablespoonful of the red water could contain about 50,000 such organisms, or 60 million per liter. In the presence of these forms, fish die in great numbers and are washed ashore. It is believed that the dinoflagellates release a poison that paralyzes the respiratory system of the fish. The result is the mass suffocation of fish in affected

American Museum of Natural History

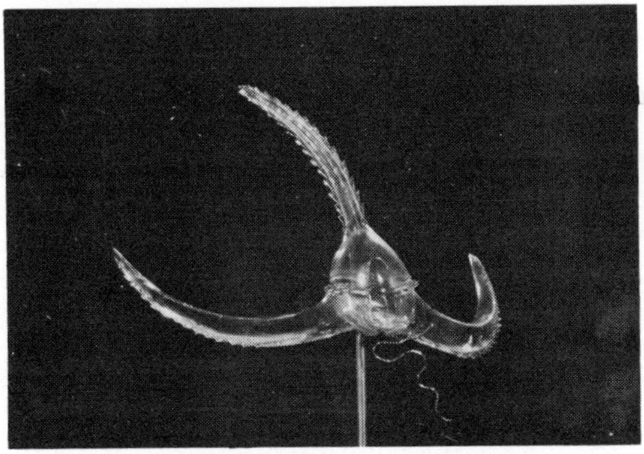

Fig. 7-3. A glass model of the luminescent dinoflagellate *Ceratium tripos*. Some phytoplankton give off short bursts of light when the water in which they living is agitated.

coastal areas. Other coastal regions are similarly affected by other species of dinoflagellates, and blooms of toxic plankton in fresh-water lakes near the sea have caused the death of water fowl and domestic animals. Humans who have breathed concentrated vapors of the afflicted waters have also developed respiratory difficulties. Such dinoflagellate blooms may produce large deposits of dead organic matter, as at Walvis Bay, South Africa; these have been considered to represent a possible source of petroleum, resulting from the bacterial reduction of the organic matter to hydrocarbons.

But some dinoflagellates (*Ceratium, Pyrodinium,* and *Noctiluca*) create the strange and fascinating light of the ocean waters—bioluminescence. There are many coastal areas in the world where permanent luminous colonies of dinoflagellates can be found. "Fire Lake," which is near Nassau in the Bahamas, and Oyster Bay near Falmouth, Jamaica, are two areas where this phenomenon occurs. Another coastal area, Phosphorescent Bay on the southwest coast of Puerto Rico contains a large population of dinoflagellates. The agitation of the water at night will cause these forms to produce enough light for a person to read by. Biological luminescence in Chesapeake Bay caused by phytoplankton has recently been measured using a sensitive light-meter apparatus. A bathyphotometer can detect flashes from luminescent organisms at the surface levels and to depths of 3750 meters (12,300 feet). As previously mentioned, some plankton are known to follow a daily rhythm of depth change that is related to intensity of solar radiation. In Chesapeake Bay the luminescent plankton were found to be concentrated at a depth of about 7½ meters (25 feet) during daylight hours. The level of maximum luminescence was found just below the surface during evening hours. Results of the light-meter experiment are probably similar to the results that would have been obtained by multiple plankton trawls in Chesapeake Bay and a counting of plankton recovered from various depths.

These, of course, are not the only organisms that possess the property of bioluminescence. Radiolarians, hydroids, jelly fishes, olcyonarians, ctenophores, gastropods, bivalves, cephalopods, protochordates, and true fishes use this property as a means of protection or as a means of food hunting. The deep-sea prawn (*Systellaspis affinis*) possesses 125 light organs on its eyestalks. It is one of the most colorful organisms in the sea. Certain fish also contain luminous organs. Two animals which possess a unique form of luminous organs are *Photoblepharon* and *Anomalops*. Both possess a cream-colored organ that is really a colony of luminous bacteria under each eye. *Anomalops* rotates this organ downward until the light is obscured by a black envelope of tissue. *Photoblepharon* draws a fold of black tissue on the lower edge of the organ socket up over the light. Other animals that possess special luminous organs are found specifically in the abyssal waters.

There is a very significant group of planktonic organisms that are classified generally on their size. They are given the name *nannoplankton*. This group comprises smaller diatoms, small dinoflagellates, coccolithophores, small protozoans, and bacteria. They are so minute that they evade the finest trawling nets. The only method of collecting them for study is by centrifuging the water sample. Larger plankton are usually strained out of sea water through fine-mesh conical-shaped nets similar to the wind socks at airports (see Fig. 7-4). One sampler that is used to determine the numbers of plankton in a volume of water has a lid that is opened and closed by "messengers." These messengers are sent down via a weighted line when the depth of trawl is reached and when the sampler is

Woods Hole Oceanographic Institution

Fig. 7-4. Scientists aboard the research vessel *Atlantis II* lower a plankton net. These fine-mesh nets are towed near the surface to obtain plankton.

to be brought to the surface. A flow meter is placed at the opening so that the volume of strained water can be measured. In this way, an uncontaminated sample of plankton can be obtained from any specific depth. Some plankton samplers allow continuous towing for hundreds of kilometers.

The *coccolithophores,* an important group of the nannoplankton, are the smallest photosynthesizing organisms of the sea. They are biflagellated and possess calcareous protecting armor. These calcified plates or shields of various designs can be found in enormous numbers in marine bottom deposits. They belong mostly to the open sea, but do reproduce in coastal waters. They can number about 5 to 6 million per liter.

which gives certain waters a milky appearance. The North Sea fishermen welcome this cloudy phenomenon since it is a sign of a good herring haul. The coccolithophores are important as food to filter-feeding organisms and contribute to the calcareous bottom sediments.

All the various kinds of diatoms, dinoflagellates, and nannoplankton, have been given the general name *grasses of the sea.* Dependent upon radiant energy for photosynthesis and upon the presence of certain inorganic nutrient substances (carbon dioxide, water, nitrates, phosphates, iron), the grasses of the sea are able to manufacture carbon compounds. The stored energy in these small plant cells becomes the source of chemical energy for all life processes. Estimates indicate that the grasses of the sea produce from 80 to 120 billion tons of carbohydrates per year, in comparison to terrestrial plant production of 40 billion tons. From the carbohydrates the phytoplankton build proteins and fats. It has been determined that these organisms contain 40–50 percent protein by weight, 20–40 percent carbohydrate, and 20–25 percent fats.

Zooplankton. Herbivorous zooplankton are the *primary consumers* or *grazers* of phytoplankton and derive protein, carbohydrate, and fat directly from these organisms. The zooplankton covers a multitude of phyla from the protozoans to the larvae of true fishes (see Fig. 7-5). The tintinnids, radiolarians, and foraminifera compose the greatest number of the protozoans in the zooplankton. The unicellular *tintinnids* swim by means of a whorl of cilia at the anterior (forward) end. They feed on the small nannoplankton. Their shells vary from tubular forms to urn-shaped structures.

The *radiolarians* are planktonic organisms whose skeletons are composed mainly of silica (see Fig. 7-6). These organisms secrete some 4400 designs and have long slender spines radiating in all directions. The length of these spines can be adjusted to the density of the water to keep the animal afloat. Upon sinking and mingling with the bottom sediments, the skeletons become the constituents of the siliceous radiolarian oozes covering the floor in the deep tropical waters of the Pacific Ocean.

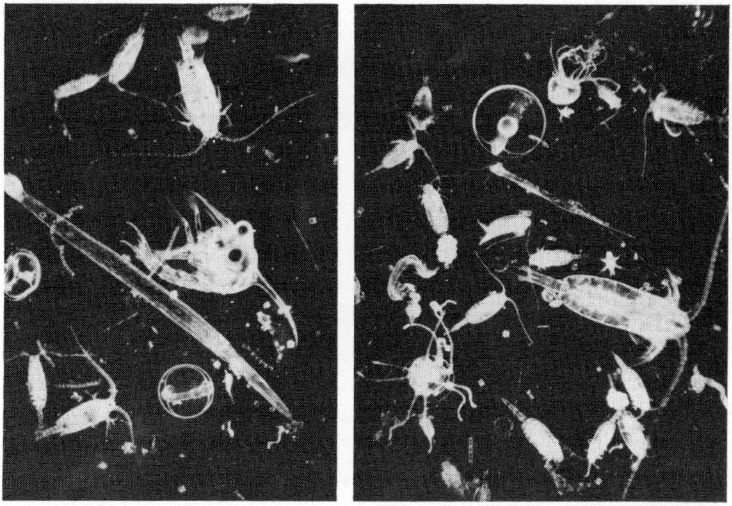

Woods Hole Oceanographic Institution

Fig. 7-5. An important link between the phytoplankton and the larger nektonic organisms in the food chain is the zooplankton. The thin elongated organism is an arrow worm. The organisms with the two large antennae are copepods. The tentacle-bearing organisms are larval forms of jelly fish.

A deep scientific interest for the order *Foraminifera* lies in the skeletal structures produced by its members. Shells are formed with one or more chambers arranged in a straight line or spiral. They are constructed of calcium carbonate, but some possess silica and chitin. Other forms possess pores through which the protoplasm appears as finger-like appendages. Upon sinking to the bottom, the shells form an important constituent of globigerina ooze, named from the genus *Globigerina.*

In his attempts to gain knowledge about the life that inhabits the sea, man has found that some marine animals give indications of the temperature of the water in which they lived. Some of these tiny temperature-indicators are the foraminifera. Marine biologists believe that two of these formas (*Globigerina pachyderma* and *Globorotalia menardii*) coil their shells to the left in cold water and to the right in warm water. The fossil shells could then reveal the temperature of

American Museum of National History

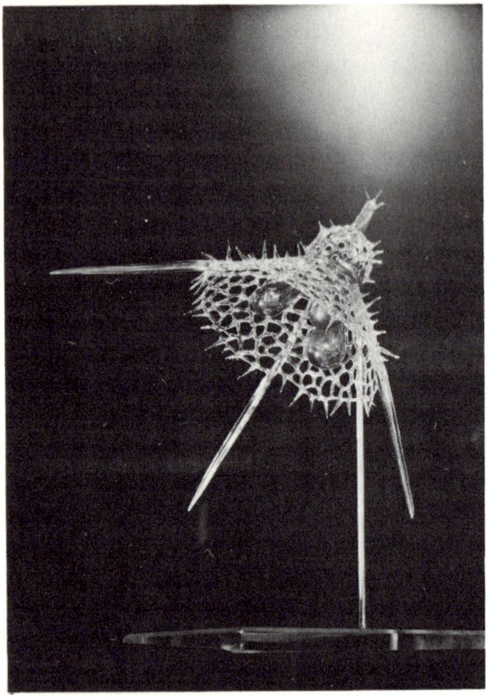

Fig. 7-6. A glass model of the radiolarian *Dictyophimus hertwigii* found in the central Pacific Ocean. Each species of radiolarians is classified according to the shape of the shell. There are at least 4000 different shells that have been recognized. Radiolarians are part of the zooplankton population and contribute to the development of sedimentary layers on the ocean floor.

ocean waters during certain earth ages. It is possible that increased knowledge of these planktonic forams may help scientists to determine the causes of ice sheets of the Pleistocene (Ice) Age. Besides the information about water temperature, some specific foraminifera lead scientists to other areas of study. The *micropaleontologist* (a scientist who studies microscopic fossils) has used the shells to guide him to potential oil-bearing rock strata.

But it is the crustaceans, comprising 70 percent of the zooplankton, that are the most numerous of the grazers of the sea. We have previously indicated the enormous numbers of these organisms in specific areas such as the Gulf of Maine. Off Bermuda 91 percent of captured plankton have been found to be crustaceans, and over 75 percent were copepods. Most copepods strain the phytoplankton from the ocean, converting their substance into animal material. In turn, they constitute an important item in the diet of many fish. The smallest copepods are about the size of a dot, and the largest are about a half-inch long. The most prominent feature is the larger anterior antennae, which may exceed the length of the animal's body. They occur in all ocean waters, but are most abundant in colder areas of the temperate zone.

The copepod *Eurytemora* is said to eat as many as 120,000 small diatoms in a day. Another important copepod, *Calanus* (see Fig. 7-7), is a link between the planktonic life and larger marine forms. When young crustaceans tend to collect in swarms during the breeding season, the appearance of blue whales (*baleen whales*) coincides with the periods of swarming. The baleen whale feeds on miniature crustaceans (*Euphausia superba*), generally called *krill,* during the winter feeding season in waters near Antarctica. It may consume as much as three tons of krill per day.

The blue whale is not the only marine organism in the Antarctic that feeds on krill. The crabeater seal, winged birds, Adélie penguins, small fishes, and squid consume large amounts of these crustaceans. The small fishes and squid become primary food sources for the larger organisms such as emperor penguin, Weddell seal, and Ross seal. The Weddell

Woods Hole Oceanographic Institution

Fig. 7-7. The copepod *Calanus* is an important intermediate organism in the food chain of oceanic life. This organism is a primary consumer of phytoplankton and is, in turn, a primary food fuel for smaller nektonic organisms, such as herring and mackerel, and larger filter-feeding animals.

seal, Ross seal, Leopard seal, and blue whale, in turn, are consumed by the killer-whale population.

Another link with larger marine forms is the herring. *Calanus*, a copepod, is a principal food of herring which consumes as many as 60,000 of these individuals during a feeding period. Mackerel, sardines, anchovies, flying fish, and the one-ton sunfish Mola Mola are also plankton filter-feeders. There must be an enormous number of copepods in the ocean if the herring and related species that feed upon these organisms are the largest measured population. They number 1,000,000,000 mature individuals over hundreds of miles of the northeast Atlantic. Hence, the destiny of the zooplankton primarily lies in the consumption by larger marine organisms. However, large numbers of zooplankton die and decompose to increase the nutrient wealth at the ocean bottom.

Nekton

Herring, mackerel, sardines, and other small forms, in turn, become necessary items in diet of larger marine animals at the surface or intermediate waters. Adult fishes, whales, squids, dolphins, seals, some crustaceans, and most swimming animals in the surface and intermediate waters are placed in the general category termed *nekton*. They swim in both waters of the *neritic* and *oceanic provinces* and possess efficient locomotive organs to swim against currents and waves. These free-swimming organisms have a tendency to form into schools and are of value as food for humans or for the abyssal-dwelling animals.

The intermediate and abyssal nektonic fishes have evolved differently from the surface-dwelling forms. Their structures are drastically altered and adapted to the increasing pressures of depth and the decreasing intensity of light. Deep-sea fish are supplied with large sharp teeth for capturing their prey. Some species can swallow other fishes much larger than themselves and store them in their inflatable stomachs. The eyes of deep-sea fish are modified in two directions: they are either very large, so as to gather the subdued light, or are very small or absent. Their bodies are very soft and the tissues are loose. Dark and silvery shades predominate, but red fins and filaments do occur. Luminous organs are present in many deep-sea forms. They appear as bright "jewels" set in the skin and are arranged in rows either along the sides or the ventral surface (underside).

Benthos

Waiting for a living animal to pass within its reach or for a dead animal to descend in its vicinity, the animal of the sea floor lives a competitive life as do the abyssal swimming forms. These bottom-dwelling forms compose the general term of sea life called the *benthos.* They dwell on the coastal floor to the deep-sea bottom and may be either sessile or mobile. In the coastal regions, the sessile forms comprise the chitons, limpets, mussels, barnacles, corals, and tubeworms that are adapted to withstand the impact of shoreline waves. The less rigidly attached animals are sea anemones, sponges, and some hydroids. The mobile animals of coastal waters include flatworms, nemerteans, annelids, starfish, crabs, scallops, and many others. They obtain their food either from other dead animals and plants, living animals and plants, or the nutrient substances in the mud of the floor. This is also true of the deep-sea benthonic forms. Sea urchins, brittle stars, sea cucumbers, snails, bivalves, and glass sponges compose a few of the deep-sea animals. They are mainly mud-dwelling. Other mud-dwelling forms that possess either elongated appendages or long-stalked branches are isopods, shrimps, sponges, hydroids, and sea lilies.

If the phytoplankton are not consumed by the zooplankton, their destiny is to decompose into organic material (see Fig. 7-8). The zooplankton, however, has received the necessary carbohydrates and energy to live and eventually be a source of food for larger predators. The zooplankton, if not eaten by predators, will also decompose into organic substances. The predators, herring, anchovies, sardines, devour enormous amounts of zooplankton in order to exist, and they, in turn,

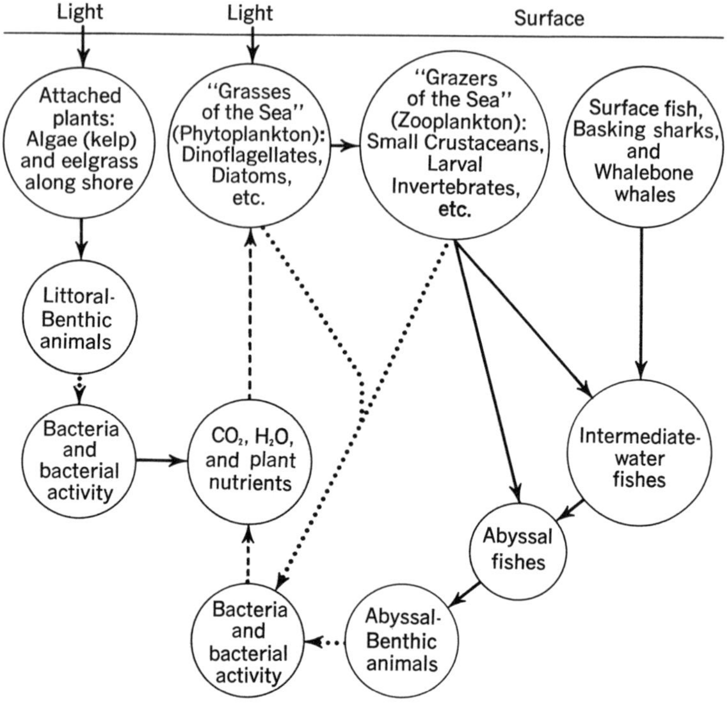

Fig. 7-8. Coastal plants, such as algae and eelgrass, also absorb light and carry on the photosynthetic process in the same manner as phytoplankton. Both plant forms become food fuel for the animal life of the sea. However, the entire food chain depends on the activity of bacterial forms. If these organisms were not present, there would be no production of nutrients that are necessary to maintain oceanic life. The solid line represents the consumption of organisms by larger predators. The dotted line represents the decomposition of plant and animal life by bacteria. The broken line represents the production of nutrients by bacteria.

become the food fuel for other nektonic organisms. Enormous numbers of these organisms die and are readily decomposed by the bacteria that dwell at all levels of the ocean. If their lot is to be food fuel, they indeed supply the larger organisms—squid, porpoises, tuna, bonito, and others—with the energy and organic substances for existence. The abyssal free-swimming animals await these large predators that may be alive or dead to descend into the abyssal waters. They have no hesitation in snatching whatever morsel descends or passes within their reach. At the abyssal level, life is extremely limited, food is scarce, and death is the only result. Eventually the abyssal free-swimming animals die and become a food supply for the bottom-dwelling animals. Benthonic animals live a more limited existence than any other oceanic form. They must compete for food that either passes them accidentally or rains upon them. Some forms have the ability to obtain their food from the ooze of the ocean floor. These benthonic animals die and are broken down to organic materials.

The "work horses" of the process of decomposition of the entire oceanic life are bacteria. They are found at all oceanic levels and have been numbered about 420,000,000 cells per grain of wet mud of the ocean floor. Some bacterial forms have been found 140 feet in mud below the ocean floor. Most bacteria of the sea are heterotrophic—they obtain their energy by the oxidation of *organic* compounds. Hence, they live as saprophytes or parasites. The phytoplankton and other algae sythesize organic substances from such inorganic raw material as carbon dioxide, nitrates, phosphates, and others. Over a long period of time organisms die at the same rate as they are born. A return of the raw material is possible if an intermediate factor exists. Heterotrophic bacteria perpetuate this process of the food cycle through mineralization of the organic material.

Fish Population and the Food Supply

Scientists and fishermen have identified approximately twenty thousand species of fish, and it has been estimated that the fisheries of the world harvest about 77 million tons of marine animal life. The gross value of this harvest approaches one billion dollars a year. There is no doubt that fishing is an eminently important industry and vital link to food resources. This aspect of food resources has placed Japan in the first place as the world's harvester of fish. Leadership in fishing productivity is essential for the Japanese national diet as well as for balance of Japan's international trade payments. Japan's fish harvest amounts to 15 billion pounds per year about three times the United States total. The United States was able to manage a second position for some years, but has been dropped to fifth place behind Japan, Peru, Communist China, and the Soviet Union. The seven most productive fishing nations and their percentage of the 1962 world fish catch are shown in Table 7-1. Preliminary tabulations indicate that the United States catch in 1963 had dropped to 4.6 billion pounds with a loss of nine million dollars relative to the previous year. Slightly less than half of the United States catch was used for meal, oil, bait, and animal food.

Fishing is the oldest industry in the United States. New England fishermen were the first to reap the plentiful harvest of fish to be found off the Grand Banks and Georges Banks in the western North Atlantic Ocean. Somehow the United States fishing fleet has tended to become less efficient because older ships have not been replaced by more modern floating fish-processing plants such as those now being operated by the Russians. Another complaint is that young people in the United States have not been attracted to either the study of fish or to fishing as a career. Our failure to recapture first place among the fishing nations of the world must be rated an expensive and avoidable national disgrace.

The species of fish harvested depends on the fish population, efficiency of the fishing fleet, industrial use, and national taste

Table 7-1. World catch of fish and other edible marine animals in 1962.

	Billion pounds (live weight)	Percent of world catch
Japan	15.1	15.3
Peru	15.1	15.3
Communist China	11.1	11.2
USSR	8.0	8.1
USA	6.5	6.6
Norway	3.0	3.0
Canada	2.5	2.5

for fish food. Presently, 125 species of fish are caught in the United States, but nearly half of the catch is accounted for by six varieties of fish (listed in Table 7-2). Most fishing grounds are located over shallow continental shelves, or where upwellings bring fresh supplies of nutrients for growth of plankton. The Grand Banks of Nova Scotia, a shallow continental shelf area, is noted for its population of cod, halibut, haddock, and redfish. In these shallow waters, sonar has proved valuable in locating schools of fish. Sound waves can tell fishermen where schools of fish are lurking and pinpoint the species and size of the individuals in the schools.

Locating a school and netting it are two different problems; the sports fisherman is able to use visual stimulus of a string lure to attract his fish, but the commercial fisherman needs a more efficient method. The commercial catch has increased because of the use of new techniques; such as electrical, magnetic, chemical, optical, and bio-acoustical lures. Menhaden trapped in a net are attracted to a bell-shaped copper end of a hose leading down from the ship. The reason for the attraction is a slight electrical charge at the net-end of the hose. The menhaden can thus be pumped aboard, through the hose, without bringing up the net.

While fishermen are searching for better and faster methods of catching fish and processing them for market, they also face the possibility of depleting some of the fisheries resources. A resource, if conserved, could help meet the great demands for food by the rapidly increasing population of the

Table 7-2. Varieties of fish that accounted for nearly half the catch by the United States fishing fleet in 1956.

Variety	Catch in Thousands of Metric Tons
Menhaden	942
Tuna	125
Salmon	123
Sea Herring	116
Haddock	69
Atlantic Ocean Perch	69
Pacific Halibut	19

world. It is imperative that conservation be practiced and that more intensive research into the habits of fish be pursued. The methods and suggestions for fish conservation are discussed in the section on the future of marine biology.

Only recently have experiments been devised to determine the mortality rates of larval and juvenile members of fish groups. Practical tags and tagging procedures are so new that the most basic information on migration is only now becoming available. The herring shoals change their encampments with the seasons, making a few long journeys (for example, between Norway and Iceland). One of the most extraordinary migrations is that of the Pacific salmon. The young are born in one of the cold rivers of the Pacific Northwest. The young fish migrate downstream to the Pacific Ocean. They eat small shrimps, arrow worms, and small fish near the continental shelves and in the open sea. They are not seen for several years. After two to six years, the salmon prepare to return to fresh water. They collect in schools and head for their native streams—in fact, to the small tributary stream in which they were spawned. This is now an established fact. The salmon fights upstream to the spawning grounds. As a result of the work of Arthur Hasler and Warren Wisky, the mechanism of salmon migration is known to involve the acute sense of smell. The salmon is guided by the distinct odor of the gravel and humus of its birthplace.

There is also the amazing migration of the American and European eels that travel thousands of miles from small tributaries to the Sargasso Sea in the western Atlantic Ocean. This is the spawning grounds for both varieties of eels. The young then return to the rivers of America and Europe. The eel's navigation cannot be explained in the present state of our knowledge.

Ocean Mammals

Man has always thought that the dinosaurs of the past were the largest animals which ever lived. As seafaring men described their encounters with large sea monsters, the thought of a larger organism was unbelievable. It is a known fact that whales are the largest animals. A blue whale can reach 100 feet in length and weigh as much as 119 tons (twice the weight of a dinosaur). Whales, porpoises, and dolphins are warm-blooded air-breathing mammals. Many whales possess rudimentary hind legs that are buried within the musculature of their body. The bottlenose dolphin (porpoise) may reach 10 or 12 feet and may weigh around 300 pounds. The spouting or blowing of a whale or porpoise is caused by the exhalation of air from the lungs, and the blowhole is no more than the external nostril. Some scientists claim that, when the blowhole is tightly closed, the whale and porpoise have stayed underwater for considerable lengths of time. The bottlenose dolphin has been recorded as diving for over seven minutes underwater.

The speeds of the larger whales have been estimated at 3 to 4 knots. Some have increased their speeds to 10 or 12 knots to avoid disturbances. The bottlenosed dolphin has been recorded as traveling at speeds of 15 miles per hour. Other scientists have estimated this mammal's speed at 22 miles per hour.

Porpoises and seals have recently become the subject of intensive research efforts. The streamlined body of the porpoise has been the subject of study by men who design submarines for the U.S. Navy. The speeds of the porpoise have been attributed to the reduced friction to the passage of water. As a result of these studies, it was discovered that the entire body of these mammals undulates in waves according to the turbulence of the water. Therefore, the external shape assumes the natural contour of the flow of water that is rushing past.

There is also the interesting possibility that porpoises are among the most intelligent of sea creatures. Some scientists have worked with porpoises and found them to be very receptive to training. Several experiments revealed the porpoise's reactions to pain and pleasure in the centers of the brain in which electrodes were attached. They were able to master these trials in less time than some species of monkeys. But difficulties stand in the way of comparing the intelligence of porpoises with that of other animals: One is the difficulty in studying them in their watery surroundings and the second is that they lack manipulative parts such as hands or fingers. Porpoises have also been tested for their sonar abilities and also for their ability to duplicate human speech sounds. They emit trains of squawks, clicks, or make creaking noises. The echoes of the sound pulses are picked by the animals after they are reflected from various objects in the water. In this way, they can locate objects, determine the distance to them, and can distinguish between different fish by the echo pattern.

The study of the anatomy and physiology of the Weddell seal may lead the scientist to discover how the human body can resist the crushing weight of water in deep dives. It is known from some experiments that the Weddell seal, which lives near the Antarctic, is able to dive to depths of at least 350 meters (1150 feet). This seal is also able to remain submerged for 10½ minutes during an average dive.

Future of Marine Biology

The increasing knowledge of planktonic life and more comprehensive observations on nektonic and benthonic life has induced the more industrious countries to investigate the harvesting of these organisms. There are numerous factors that hamper such developments. One of the limitations in harvesting is the enormous amount of sea water that must be strained to obtain an adequate amount of plankton. It is estimated that a North Sea Fisherman in a good year can harvest about 58 tons of herring per hundred hours. To gather an equivalent amount of plankton, it would be necessary to strain about 58 million tons of water. However, it may be possible to develop plankton "farms" near coastal areas in which large planktonic patches may accumulate. Recent experiments have shown that human consumption of plankton proved beneficial, and no sickening effects were observed. Some consumers have claimed that planktonic food tastes like shrimp, lobster, and vegetables. Others stated that it had a rather pleasant taste with a touch of the odor of violets. The Japanese have been farming specific plankton algae in shallow ponds containing nutrients on which plankton could exist. They have centrifuged these organisms and processed them to a dry powder. The powder has been used as a protein supplement in flour.

Kelp and seaweed have been successfully used in rice-diet areas. Actually, there is a large harvest of seaweed in the Pacific coastal areas, both in Japan and the U.S. These large algae have been also used as a source of nitrates. Furthermore, some algae have been a useful ingredient of low-calorie bakery products prepared for reducing diets. Jelly products from some red algae are being used as preservatives in canned mushrooms, ice cream, malted milk, and cheese. The larger algae and seaweeds will prove more valuable as the scientist investigates their biochemical functions. They may become a necessary item in the food consumption of heavily populated areas.

From the estimates of the population of the herring family alone and other related species, it is incomprehensible to envision an era when the human population could consume the known animal life of the sea. The present estimates on the world's total catch of fish is a drop in the oceanic bucket. If there were the remotest possibility of such a consumption, man would still have a source of oceanic food. In intermediate and deep waters, certain fish, such as *Cyclothone,* have been estimated to surpass the herring family in population. It can be used to provide raw material for a fish-protein concentrate or "fish flour." Other deep-dwelling organisms such as rat-tails (*Malacocephalus*) are not large enough for direct consumption, but could be used for fish meal or flour. They also have populations surpassing the herring family.

The methods for future fishing are impressive and promising. They cover many realms of science—from farming to electricity. In the Caspian Sea, the Russians have used electric lights to attract fish and then scoop these organisms up with a suction pump. This process seems possible in fresh-water areas such as the Caspian Sea. But in sea water, which is a good conductor of electricity, this method is rather hazardous. West German scientists have developed an electronic device that upsets the organism's normal escape reactions, and they are easier to net.

Recent experiments have proven the value of large-scale transplanting. With this method, young fish would be transplanted to specific areas that possess sufficient nutrients but not enough animal life. This would enhance fishing for areas that are located near such "fish farms." The striped bass, shad, and soft-shell clams from the North American east coast have been transplanted successfully to the Pacific shores. The North American Chinook salmon has been transplanted to New Zealand with favorable results.

There are tentative plans to mine the sea bottom with huge vacuums. Since the sea bottom contains valuable amounts of manganese nodules and other elements, these tentative methods seem very promising. They also could be promising to the marine biologist. He would be able to collect specimens and marine food from the bottom by these techniques. This vacuum technique may also overlap into the fishing industry—crabs, lobster and other bottom-dwelling edible organisms could be harvested.

8

Research Submersibles

Men have always been drawn to the alien world beneath the ocean's surface. Their purposes may have been for direct observations, for mere curiosity, or for economic reasons. But man's physical structure imposes certain limitations on the depth to which he may proceed. Those who have made free dives to the shallow bottom of a swimming pool, lake, or just outside the breaker zone at the seashore, know the uncomfortable feeling caused by pressure on their bodies. At depths of 3 meters (10 feet) our ears "pop" to equalize the pressure on our eardrums. In addition to the effect of pressure on our ears, there are many other physical effects. Without a mask, for instance, a diver will experience a rapid loss of visual acuity caused by pressure on his eyes. If care is not exercised, his sinus cavities will fill with water. For this reason, sinus infections are common among springboard divers and inexperienced novices. There are, however, exceptions to man's limitations of descent into the depths. Island people of the tropical Pacific Ocean have made free descents to about 46 meters (150 feet) in search for pearls.

In addition to the physical limitations on free descents, free ascents to the surface also affect man. The human body must compensate for the sudden decrease in pressure that it encounters during ascent. If the pressure decrease is faster than the rate of compensation, the effect will be more serious to the diver's body. Since this problem had been experienced during emergency ascents from submarines, the U.S. Navy has developed specific projects concerning free ascents. Commander George Bond, a U.S. Navy medical research officer, has made ascents from the emergency hatch of a submarine to test human endurance. These tests reached a marvelous culmination on October 2, 1960, when Commander Bond ascended from 91 meters (300 feet) without auxiliary breathing apparatus.

Many devices have been developed to give man more freedom in the oceans and to help him overcome his physical limitations. One of the earliest references to free diving is contained in Homer's *Iliad.* Aristotle stated that divers were sometimes provided with breathing tubes for drawing air from the surface. This device was useful only in the upper few meters of water because of the pressure on the diver's body. Aristotle also referred to a metallic vessel that was filled with air and lowered to divers as an auxiliary air supply. Alexander the Great was recorded as making descents in a device called a *colimpha* that allowed a diver to remain dry during the descent. This device is usually pictured as being a glass vessel because it was said that the diver could see the watery surroundings. However, out of deference to other Greek engineering achievements of the same age, one might speculate that the colimpha was metallic and was fitted with one or more watertight portholes. In any case, this device is considered the forerunner of the modern diving bell.

Two thousand years had passed before an English astronomer, Sir Edward Halley (better known for discovering a comet), invented an apparatus to carry air from the surface down to the diving bell. This invention consisted of two large barrels filled with air and attached to the diving bell. Each barrel had an open hole in the bottom and an air hose extending from the top. So long as the air hose hung lower than the holes on the bottom, the water pressure prevented the air from escaping. To transfer air from the barrel to the inside of the bell, the diver lifted the open end of the hose to the top of the bell, and the air flowed in. The continued raising and lowering of the barrels supplied the divers with fresh air. This invention did not allow the diver enough maneuverability and time in the water. It also limited the depth of the descent.

Man continued to probe the unknown with many unique and bizarre devices. It was not until 1791 that a tin helmet with pipes leading to the surface was invented and thus established the first prototype of the "hard hat." This was the first use of air forced down through a pipe to the helmet. In 1837, the invention of the closed diving suit with both intake and outlet valves in the helmet revolutionized diving techniques. This device, the "hard hat," became the model for all future helmet-suits. This 19th century invention gave man the ability to remain at great depths for a reasonably long time and allowed freedom of action in this environment. At present a hard-hat diver can operate efficiently to a maximum depth of about 185 meters (610 feet).

The history of independent underwater-breathing apparatus began about 1879 when an English merchant seaman carried a tank of pure oxygen on his back during dives. Two breathing tubes extended from the tank to a rubber mask on his face. This device was safe to use only in shallow water. Below 30 feet, the oxygen in the tank caused convulsions and even death under pressure.

In 1943 Jacques-Yves Cousteau improved this apparatus. He and the French engineer Emile Gagnan devised a fully automatic regulator that would feed the proper amount of air to the diver regardless of his position. This became the now-famous Aqualung or *Scuba* (self-contained underwater breathing apparatus).

The Bathysphere

Descents to greater depths were facilitated by Otis Barton's development of the steel diving sphere in 1930. With financial support from the New York Zoological Society and the National Geographic Society, a steel sphere, large enough to accommodate two men, was designed. Its internal diameter was 1.37 meters (4.5 feet) with a wall thickness of 3.8 centimeters (1.5 inches). The sphere weighed 2273 kilograms (5000 pounds) in air. There were five viewing windows with diameters of 15 centimeters (6 inches) that were covered by fused quartz glass. The glass was 7.5 centimeters (3 inches) thick. This diving chamber was called a *bathysphere,* which means "deep sphere" (Fig. 8-1). The air supply for the vehicle

New York Zoological Society

Fig. 8-1. Dr. William Beebe and Otis Barton after their 1009-meter (3028-foot) dive in the bathysphere.

was self-contained, power and communication were maintained through an electrical cable between the bathysphere and the mother ship. At Nonsuch Island, Bermuda, the bathysphere was tested on 31 preliminary dives. On August 15, 1934, Otis Barton and William Beebe reached the shallow bottom at a depth of 923 meters (3028 feet). The descent time was 1 hour and 14 minutes, and the vessel was submerged for 3 minutes at that level. The ascent took 1 hour and 23 minutes. This depth record remained until 1948, when Barton descended in the *benthoscope*, a vessel similar to the bathysphere. On this occasion the descent was 1372 meters (4500 feet) off the California coast.

The development of research submersibles continued with the work of the Swiss scientist, Auguste Piccard, who was a mechanical engineer, doctor of natural sciences, and professor of aeronautics at the Polytechnic Institute in Brussels. He sought to become the "Columbus of the atmosphere," and so he designed an airtight aluminum gondola supported by a hydrogen-filled balloon by which he hoped to ascend higher than previous explorers. On May 27, 1931, Dr. Piccard and an assistant rose from Augsburg, Germany, and reached an altitude of 15,791 meters (51,775 feet). They descended near Obergurgl, Austria, to world acclaim for this ascent. Piccard was not content with only the altitude record to his credit, and so he decided to descend to greater ocean depths than any previous scientist. He and his son, Jacques, with a small group of engineers and scientists designed and built a deep-diving vehicle. The *bathyscaph* (deep boat) was the first deep-diving vehicle with both self-contained electrical power and air supply and, hence, was not physically connected to a surface ship. This vehicle, after being tested in shallow waters, was acquired by the French Navy and renamed *FRNS-3*. In 1954, *FRNS-3* made a record dive to 4050 meters (13,284 feet) off Dakar, West Africa.

Trieste

Not idle during this period, Auguste and Jacques Piccard in 1952 worked on an improved bathyscaph that they named *Trieste* (see Fig. 8-2). After successful preliminary tests, this vehicle was purchased by the U.S. Navy Electronics Laboratory in 1958 and continued to prepare for dives into the deepest

Official U.S. Navy Photograph

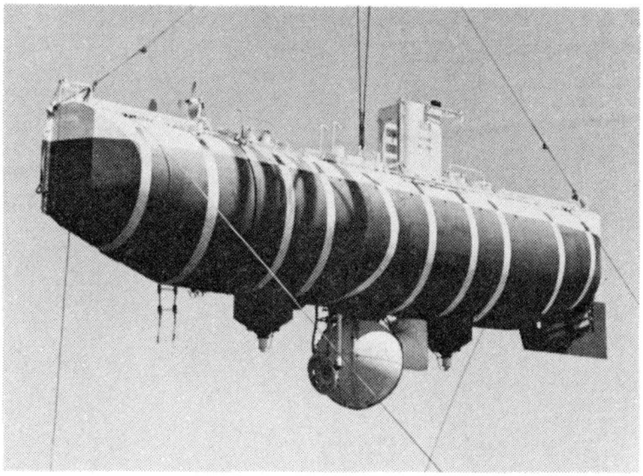

Fig. 8-2. The U.S. Navy research bathyscaph *Trieste I* being lowered into the water after an overhaul performed prior to the deep dives of "Project Nekton." The *Trieste* set a depth record during this project.

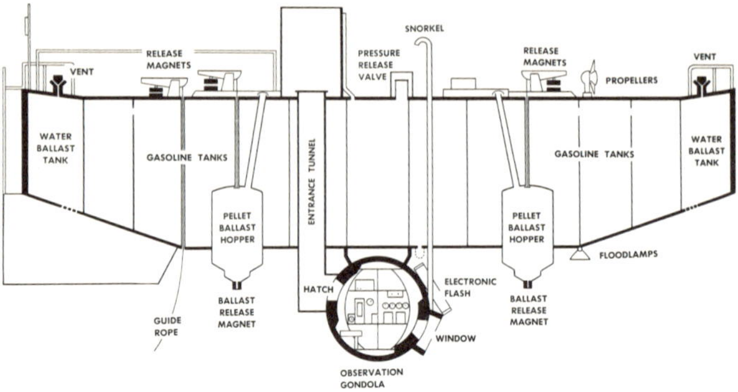

Official U.S. Navy Photograph

Fig. 8-3. Diagram of the interior of the *Trieste I*, showing the location of personnel gondola, specific scientific equipment, and the ballast tubs.

parts of the ocean—the oceanic trenches. On January 23, 1960, Jacques Piccard and Lieutenant Don Walsh descended into the Mariana Trench (Challenger Deep) near Guam in the Pacific Ocean (Fig. 8-3). The bottom of the trench was recorded at an official depth of 10,915 meters (35,800 feet). Descent time was 4½ hours. Piccard and Walsh observed the bottom for about 20 minutes and, then, because of a weak plexiglass porthole, ascended to the surface in only 3½ hours. Even though the plexiglass porthole had cracked under the extreme pressure (approximately 1197 kilograms per square centimeter or 17,000 pounds per square inch), it was later determined that the scientists had not been in any real danger.

Trieste is really an underwater balloon in which gasoline keeps the vehicle afloat (see Fig. 8-4). For the record-breaking dives of the U.S. Navy's "Project Nekton" in 1960, the gasoline float mechanism was increased in length to 17.7 meters (58 feet), and the gasoline capacity was thereby increased to 130,000 liters (34,300 gallons). The two-man spherical gondola that is supported by the float mechanism is made of an 8.9-centimeter thick (3.5-inch) steel alloy. The internal diameter of the gondola is 1.94 meters (6.4 feet), and the gondola weighs 11,812 kilograms (26,000 pounds) in air. For the deepest dives, the total capacity of the two ballast tubes was increased to 14,545 kilograms (32,000 pounds) of steel shot. For *Trieste* to descend, the two water ballast tanks are filled with sea water. As descent proceeds, the gasoline, which is more compressible than sea water, loses some of its buoyancy. Some of the steel shot must then be released in order to maintain a uniform rate of descent. Small propellers that are driven by electric motors allow some maneuverability of the submerged vehicle. As there is no physical connection between *Trieste* and the surface ship, a foolproof system for ballast release was designed in cases of emergency. This ballast system is operated on the fail-safe principle. Electromagnets hold both the steel-shot ballast and the steel ballast tubs. If the electrical system fails and the magnets lose their pull, the ballast tubs are released automatically, and the vehicle rises. During an overhaul in 1961, *Trieste* was almost

Official U.S. Navy Photograph

Fig. 8-4. The ocean floor at 5673 meters (18,600 feet) in a shallow portion of the Mariana Trench. Photograph was taken through one of the 20-centimeter (8-inch) thick plexiglass portholes in the observation gondola of the *Trieste*. Rocks in the background are about 15 meters (50 feet) away.

completely rebuilt with American standard parts. The electric power system is 24 volts, with pressure compensated batteries mounted externally. Five motors provide vertical, horizontal, and transverse movement. Many new controls and instruments such as an underwater acoustic telephone, directional gyro-compass, tape recorders, scanning sonar, closed-circuit TV, and flood lights greatly improve the efficiency of the craft. The specialized scientific instrument capacity has been greatly increased. External and internal cameras with synchronized strobe lights, a mechanical arm, plankton samplers, and automatic water sampling bottles are a few of the scientific instruments. In addition, many specialized devices can be mounted, such as sediment sound velocity probes, current meters, temperature sensing systems, and oxygen monitors.

On April 10, 1963 the U.S. atomic submarine *Thresher* sank 220 miles east of Boston in 2800 meters (9200 feet) of water.

Official U.S. Navy Photograph

Fig. 8-5. The U.S. Navy research bathyscaph *Trieste II.*

Trieste was shipped from California to investigate any possibility of discovering wreckage in the area. Portions of the ill-fated submarine were finally discovered by searchers aboard *Trieste*.

The present and future program for *Trieste I* and the new *Trieste II* (see Fig. 8-5) will be to study problems in acoustics, physical oceanography, and sea-floor topography off the Southern California Coast.

A Mesoscaph

To add the final touch to the Piccard imagination, an underwater helicopter has been planned—a submersible with two battery-driven propellers on top that would drive it to depths of about 2000 meters (6500 feet). This *mesoscaph* (middle-depth ship) will be a large lighter-than-water bubble or cabin made out of plexiglass. If the propeller motors are shut off or fail to work, the mesoscaph would rise to the surface by itself. An aluminum and plexiglass housing would be mounted on top of the bubble and would contain a gasoline engine and normal propeller to drive the vehicle horizontally. This vehicle will have greater maneuverability than most submersibles and will be relatively inexpensive to build.

Denise

Captain Jacques-Yves Cousteau, known for his oceanographic research and co-development of *Scuba* (self-contained underwater breathing apparatus), supervised the construction of a two-man diving saucer called *Denise* (see Fig. 8-6). This vehicle is more maneuverable than any manned submersible vehicle in existence, but is presently limited to investigation of the continental shelf regions. The maximum operational depth of the vehicle is 300 meters (985 feet), but deeper diving vehicle of this design are considered feasible. *Denise* consists of a water-tight personnel sphere having a wall thickness of two centimeters (three-fourths of an inch). It has enough

Fig. 8-6. *Denise,* the diving saucer being exhibited on the main hatch of Captain Jacques-Yves Costeau's research vessel *Calypso.* When not in use, the diving saucer is lowered by a hydraulic crane into the after-hold.

internal space for two men to lie prone during submerged operations. The water-tight compartment is surrounded by an equatorial bulge containing batteries for underwater lights and the operation of the hydraulic system. Pumps force water through the twin nozzles of the hydrojet system. Located at the forward edge of the equatorial bulge, the nozzles can be rotated to propel the vehicle in any direction at speeds up to 0.8 meter per second (1.5 knots). The 23-kilogram (50-pound) ballast that can bring the saucer to a desired depth consists of two steel bars. These bars are attached to the hull by electromagnets. As in the bathyscaph *Trieste,* if the electrical power on *Denise* fails, the ballast is automatically released.

With the cooperation of Jacques-Yves Cousteau, Westinghouse is building a self-propelled deep-sea vehicle called *Deepstar.* This vehicle will operate at a depth of 4000 meters (13,000 feet). It is spherical in shape and will weigh seven tons. The hull will be six feet in diameter and about 1¼ inches thick. The crew will consist of an observer, a pilot who will be in a prone position, and a copilot. The *Deepstar* will have a 24-hour maximum operable time beneath the surface. The range of the vehicle will be about 20 nautical miles with speeds up to 3 knots. The craft will perform such underwater operations as the placement, observation, and recovery of instruments and equipment in buoys and vertical arrays; photography; salvage operations; sampling and coring; and advanced work on sensors, underwater lighting, and sonar equipment.

There are literally hundreds of manned and unmanned research submersibles in various stages of planning and production. Two of the most recently completed submersibles, *Alvin* and *Aluminaut* are described below.

Alvin

Alvin is a 6.7-meter (22-foot) long, two-man research submarine than can reach a maximum operable depth of 1800 meters (6000 feet). Because of this depth limitation it is intended primarily for exploration of the continental shelf and the shallow upper portions of the continental slope. There are five portholes of which four are placed around the personnel sphere for observation of marine life and geology of these regions (see Fig. 8-7). A mechanical arm extends forward from the prow of the vessel and will collect geological specimens (outcropping rocks, manganese nodules, corals) that

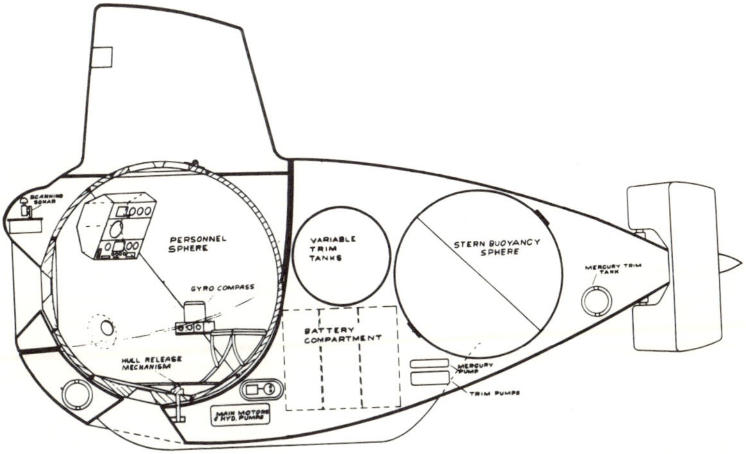

Litton Industries

Fig. 8-7. Simplified cross section of *Alvin*. In an extreme emergency, the personnel sphere forward hull assembly can be detached from the submarine body and will float to the surface.

attract the oceanographer's attention. A combination of electrical and hydraulic motors allow the vessel to cruise along the bottom at 2½ knots for a maximum of 10 hours. If necessary, short bursts of speed of 10 knots can be obtained. These bursts might be needed to move away from curious or affectionate large marine animals or to avoid avalanches of sediment from steep slopes. A burst of speed on these occasions may represent the difference between burial and survival. An additional safety feature is the buoyancy of the forward hull assembly and personnel sphere. In an extreme emergency, this portion of the vehicle can be mechanically disconnected from the remaining hull. It will immediately float to the surface. Another feature of *Alvin* is the forward sonar unit that allows the observer to identify distant obstacles and to facilitate navigation in murky water. The Office of Naval Research financed the construction of *Alvin* as an important oceanographic tool for use by Woods Hole Oceanographic Institution.

The *Aluminaut*

The *Aluminaut* is the first submarine to be built entirely of high-strength aluminum alloys (see Fig. 8-8). Its name is derived from the alloys used in its construction and from the Greek word for "diver." The vehicle was built by Electric Boat Division of General Dynamics Corporation.

Eleven huge aluminum cylinders were forged and machined to fine tolerance so that they could be bolted together to form a complete watertight cylindrical hull that has no gaskets. Each individual cylinder is 1.016 meters (40 inches) wide, 16.5 centimeters (6½ inches) thick, and 2.4 meters (8 feet) outside diameter.

The vehicle can withstand a pressure of 60,000 pounds per square inch. With the hemispherical and stern sections attached, the *Aluminaut* is 15.5 meters (50 feet, 11 inches) long. The vehicle displaces 150,000 pounds. A horizontal propeller on the top of the vessel is used for the descent and ascent. Two vertically-mounted propellers near the stern can drive the vehicle at speeds up to 2 meters per second (3.8 knots). Stor-

Reynolds Metals Company

Fig. 8-8. The *Aluminaut*, world's first all-aluminum deep-diving research submarine.

age batteries will supply electrical power for up to 72 hours of submerged cruising. Relative to the shallow working depth of other self-powered submersibles, the *Aluminaut* has an impressive working depth of 4572 meters (15,000 feet). The pilot and two scientists will have at their disposal the most modern underwater equipment, including sonar, television, and a mechanical arm for retrieving samples from the ocean floor.

These designs are strikingly similar in that each manned submersible depends upon encapsulating the observer in a thick-shelled and heavy pressure-resistant housing. It would seem that a significant breakthrough in this problem could be based on a medical accomplishment. Deep dives in pressurized vehicles are impractical because of the excessive decompression time. This duration is necessary to avoid "bends" that result from a rapid ascent to the surface. Perhaps an exotic mixture of gases could be developed that would eliminate decompression problems and thus decrease adverse secondary effects on the human body. Internal pressurization of the vehicle would allow design of less massive personnel compartments. More maneuverable vehicles could then be produced. There has been a suggestion that medical research may develop a surgical and mechanical system by which the human body could accommodate to the great pressures. Further mechanical, chemical, and medical progress might allow the diver to extract the dissolved oxygen in the sea water to satisfy his respiratory needs.

Special Vessels and Devices

In addition to the research submersibles for direct observations at the bottom, many vehicles are being designed to satisfy special needs. With financial support from the U.S. Navy, the Marine Physical Laboratory of the Scripps Institution of Oceanography has designed a most unusual research vessel designated *FLIP* (Floating Instrument Platform). It has a 93-meter (305-foot) tubular section, varying in diameter from 20 feet to 12½ feet in its submerged portion. A 15-meter (50-foot) prow that contains living quarters, laboratories, and machinery in a four-story arrangement is attached to the tubular section. For operation *FLIP* is towed to a research area in a horizontal position. The ballast tanks are flooded, and the vessel then flips into a vertical position (see Fig. 8-9). At the completion of the program, the water in the ballast tanks is forced out with compressed air, and the vessel flips back to a horizontal position. Its function is to serve as a stable and noise-free manned platform for delicate underwater acoustical observations.

Even though we are constantly achieving a greater capability to work and live in the underwater realm, there are operations too repetitive or too dangerous for divers or manned submersibles. Therefore, scientists have applied either old techniques used only on land or newly devised machines to facilitate these operations.

Underwater television achieved usefulness as a result of a tragic naval accident. In 1951, the British submarine *Affray* was sunk in the English Channel in 280 feet of water. With

Scripps Institution of Oceanography

Fig. 8-9. *FLIP,* a unique research vessel operated by scientists from the Scripps Institution of Oceanography, is free to float with the ocean-surface current while the researchers sample the physical and chemical characteristics of sea water.

the large amounts of wreckage in this area, salvage and sonar were useless in locating the wreck. With underwater television, the British Navy was able to identify the vessel and the surrounding area. Today, underwater television is an essential item in developed submersibles. Where it is dangerous for man to descend and observe underwater, television will be a vital instrument in future studies.

Where sample collections or manipulation of instruments are required, television is useless. An experienced diver must be sent down. The diver is limited to a practical working depth of 90 meters (300 feet), and while on the bottom, he needs many surface personnel to check his lights, communication, air supply, and his general safety. Advances in the development of controlled devices have aided scientists as well as salvage crews in eliminating this problem. One particular device, designated *RUM* (Remote Underwater Manipulator), allows the scientist to observe the sea floor through television cameras and will collect geological samples and marine organisms.

It also services permanent deep-sea instruments. The important feature of this device is that the scientist will control the entire operation from a surface ship or from shore.

A robot designed for underwater operations might overcome the depth limitation, shortness of diving time, and the cost of human operators. With consideration of economic and human factors, Shell Oil Company scientists designed and built the first underwater robot. The first of their creative efforts was *Mr. Mobot,* a 4.3-meter (14-foot) tall and 1.5-meter (5-foot) wide automaton. This first generation underwater robot (Fig. 8-10) has worked on wellhead adjustments in 300 meters (1000 feet) of water. There is no theoretical limit to its depth capability. *Mr. Mobot* has a television camera for eyes and a battery of underwater lights. A shipboard operator can

Shell Oil Co.

Fig. 8-10. *Mr. Mobot* is lowered into the sea from a barge during tests off the California coast.

Shell Oil Co.

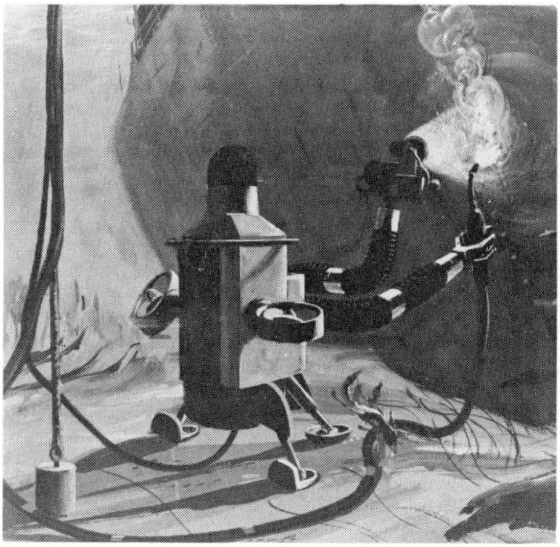

Fig. 8-11. *Mr. Unumo* using a cutting torch in an underwater salvage operation.

observe with a television receiver screen approximately 9 meters (30 feet) in front of the robot in clear water. Two adjustable propellers push the robot toward the designated work area. Mobot's ears are a sonar pinger and a hydrophone system that allow the operator to find objects on the sea floor about 425 meters (1400 feet) away. A variety of terminal appendages can be bolted onto its stubby arms as a substitute for human hands. These include a socket wrench, pincher claws, and a valve wrench. At present, these appendages can be operated only in a horizontal plane. Because of this limitation, underwater wellhead equipment was redesigned to accommodate *Mr. Mobot's* mechanical ability.

Based on the laurels of *Mr. Mobot's* success, Shell scientists anticipate the construction of a second generation underwater robot. It will be called *Mr. Unumo* (Universal Underwater Mobot) (Fig. 8-11). The robot will be a much more versatile underwater oil-field worker. Tripod feet, directional propellers, and flexible arms with hand-like appendages will allow the robot to perform almost every task that now requires the use of skilled divers. Hughes Aircraft Corporation is building an advanced prototype of *Mr. Unumo* for Shell's underwater operations.

These submersibles (vessels and robots) are the creations of humans to compensate for their limitations in the ocean depths. Humans with these devices can observe and collect whatever is necessary at forbidding depths. Jacques-Yves Cousteau and Navy scientists independently have devised methods by which humans can dwell in undersea homes and laboratories for days. Cousteau's experiments on this subject in 1963 involved five people who lived 30 feet below the surface of the Red Sea for four weeks. They lived in an underwater village in which the houses were chambers of reinforced steel, and one chamber had several rooms shooting off from a central hall. These people had the same modern conveniences found in homes on land. Laboratories for direct experiments were also included.

The U.S. Navy has recently completed experiments in the Atlantic Ocean with a 40-foot cylinder called *Sealab-I.* This project was under the direction of Commander George Bond. In this laboratory, four aquanauts lived for 10 days at a depth of 192 feet. The experiment proved that humans could live for long periods at a pressure of 86 pounds per square inch (about six times the atmosphere pressure at sea level). This permits immediate observation and sampling of oceanic material. Most important, these two experiments may answer the question of whether or not man can live permanently in the sea.

Bibliography

Books of General Interest:

Bascom, W. *A Hole in the Bottom of the Sea.* Garden City: Doubleday, 1961.
Carson, R. L. *The Sea Around Us.* New York: Oxford University Press, 1961.
Coker, R. E. *This Great and Wide Sea.* New York: Harper & Row, 1962.
Cowen, R. C. *Frontiers of the Sea.* Garden City: Doubleday, 1960.
Deacon, G. E. R. (Editor). *Seas, Maps, and Men.* Garden City: Doubleday, 1962.
Dietz, R., and J. Piccard. *Seven Miles Down.* New York: Putnam, 1961.
Ericson, D. B., and G. Wellin. *The Deep and the Past.* New York: Alfred A. Knopf, 1964.
Omanney, F. D. *The Ocean.* New York: Oxford University Press, 1949.
Shepard, F. P. *The Earth Beneath the Sea.* Baltimore: The Johns Hopkins Press, 1959.
Stewart, H. B., Jr. *The Global Sea.* Princeton: D. Van Nostrand, 1964.

Selected Articles from *Scientific American:*

Animals of the Abyss, November 1957.
Animals of the Bottom, July 1952.
Food from the Sea, October 1949.
Micropaleontology, July 1962.
Minerals on the Ocean Floor, December 1960.
Submarine Canyons, April 1949.
The Bathyscaph, April 1958.
The Circulation of the Abyss, July 1958.
The Circulation of the Ocean, September 1955.
The Coelacanth, December 1955.

Continental Drift, April 1963.
The Continental Shelf, March 1955.
The Cromwell Current, April 1961.
The East Pacific Rise, December 1961.
The Interior of the Earth, September 1955.
The Mohole, April 1959.
The Oceanic Life of the Antarctic, September 1962.
The Origin of Submarine Canyons, August 1956.
The Origin of the Earth, October 1952.
The Rift in the Ocean Floor, October 1960.
The Sea's Deep Scattering Layers, August 1962.
The Trenches of the Pacific, November 1955.
The Voyage of the *Challenger*, May 1953.
Volcanoes, November 1951.
Whales, Plankton, and Man, January 1958.

Advanced Technical Books:

Barnes, H. *Oceanography and Marine Biology.* New York: Macmillan, 1959.

Defant, A. *Physical Oceanography* (2 volumes). New York: Pergamon Press, 1961.

Dietrich; G. *General Oceanography.* New York: John Wiley-Interscience, 1963.

Heezen, B. C., M. Ewing, and M. Tharp. *Floors of the Oceans, I: The North Atlantic.* New York: Geological Society of America (Special Paper 65), 1959.

Hill, M. N. (Editor). *The Sea: Ideas and Observations* (3 volumes). New York: John Wiley-Interscience, 1962.

Huxley, A. *Standard Encyclopedia of the World's Oceans and Islands.* New York: Putnam, 1962.

Kuenen, P. *Marine Geology.* New York: John Wiley & Sons, 1950.

Menard, H. W. *Marine Geology of the Pacific.* New York: McGraw-Hill, 1964.

Moore, H. B. *Marine Ecology.* New York: John Wiley & Sons, 1958.

Nicol, J. A. *The Biology of Marine Animals.* New York: John Wiley-Interscience, 1960.

Raymont, J. E. *Plankton and Productivity in the Oceans.* New York: Macmillan, 1963.

Sears, M. (Editor). *Oceanography.* Washington: American Association for the Advancement of Science, 1961.

Stommel, H. *The Gulf Stream.* Berkeley: The University of California Press, 1958.

Glossary

Abyssal hill. Generally oval-shaped hill found in great numbers between the continental margin and mid-ocean ridge.

Abyssal plain. Extremely flat portion of the deep ocean-basin thought to result from deposition of terrigenous sediment.

Abyssal zone. Oceanic depth zone extending from 2 to 6 kilometers below the ocean surface.

Aphotic zone. Ocean depths below the lowest level at which photosynthesis can take place.

Azoic. Meaning without animal life.

Bathyal zone. Oceanic depth zone extending from 0.2 to 2 kilometers below the ocean surface.

Bathymetric chart. Map that shows the topography of some portion of the ocean floor.

Bathyphotometer. A sensitive light meter that is used to measure amount of light emitted by bioluminescent marine animals.

Bathyscaph. Literally a "deep boat," refers to any one of several research submersibles that employ a flotation compartment with a personnel sphere attached to its underside.

Bathysphere. Two-man diving sphere used by Otis Barton and William Beebe in the first-recorded deep dive.

Bathythermograph. Instrument towed behind the research vessel which records temperature versus depth on a smoked glass slide.

Bed load. That portion of stream-transported sediment that moves along the floor of the channel.

Benthos. General term for marine life found in or on bottom sediment.

Benthoscope. A diving sphere similar in design to the bathysphere.

Bioluminescence. Visible light emitted by some dinoflagellates and other marine animals.

Bottom water. Water mass moving in contact with the ocean floor.

Calcareous ooze. Ocean-basin sediment composed of the microscopic remains of marine animals that secrete a test or shell of calcium carbonate.

Coccolithophore. Member of the nannoplankton which possesses an outer armor of calcareous plates called coccoliths.

Concretion. Roughly spherical mass of material deposited by precipitation from solution.

Consolidation. Process by which sediments are compacted and water contained in the pore spaces is expelled.

Continental drift. The hypothesis that continental masses are capable of moving relative to one another.

Continental margin. That major portion of oceanic topography lying nearest to the continents.

Continental rise. Gently seaward-sloping topography lying at the base of the continental slope.

Continental shelf. Gently seaward-sloping surface lying between the shoreline and the continental slope.

Continental slope. Steeply sloping surface lying seaward of the continental shelf.

Contour line. Line connecting points on the earth's surface having the same height above sea level.

Convection-current theory. Concept that slow vertical circulation cells in the mantle are responsible for growth of mid-ocean ridges, marginal trenches, and other aspects of oceanic topography.

Copepod. Any one of a group of tiny marine crustaceans.

Corer. Pipe-like device which is driven vertically into oceanic sediment to obtain a sample.

Coriolis force. Deflecting force, created by the earth's rotation, which acts on all moving particles.

Countercurrent. A subsurface current flowing in essentially opposite direction to the surface current. Also refers to surface current such as the Equatorial Countercurrent in both the Atlantic and Pacific Oceans.

Deep scattering layer. Sonic reflecting horizon above the ocean floor believed to be composed of relatively dense concentrations of zooplankton and other marine animals.

Deep water. Water mass found lying above the bottom water mass and below the intermediate water mass.

Density. Weight per unit volume of material.

Density current. (See turbidity current.)

Diatom. Microscopic, unicellular planktonic marine plant.

Dinoflagellate. A zooplankton having whip-like processes of the cytoplasm that are used for partial locomotion.

Disphotic zone. Dimly lighted ocean depth zone lying below the euphotic zone.

Diving saucer. Two-man research submersible limited to shallow depths which is powered by hydraulic jets.

Earthquake focus. The place in the earth's crust or mantle where an earthquake has occurred.

Echo sounder. Electronic device used to determine and record water depth in the ocean.

Epicenter. Point on the earth's surface directly above the earthquake focus.

Eulittoral zone. Depth zone extending from high-tide line down to the maximum depth of holdfast plants.

Euphotic zone. Upper level of the ocean in which sunlight penetration is sufficient for photosynthesis.

Expanding-earth theory. Concept that the earth is expanding and that new crust is being formed in the mid-ocean ridges.

Fathom. Unit of depth equal to 1.83 meters (6.0 feet).

Fault. A rock fracture along which relative movement of the fault blocks has taken place.

FLIP. Acronym for floating instrument platform.

Focus. (See earthquake focus.)

Foraminiferan. Member of an order of protozoans that form chambered shells primarily of calcium carbonate.

Fracture zone. Roughly linear fault zone in oceanic crustal rock.

Geological dredge sampler. A type of sampler dragged along the ocean bottom to collect large rock fragments and nodules.

Geophone. Microphone-like instrument used to detect seismic waves generated by man-made explosions.

Graded bed. Layer of sediment in which coarse sizes predominate near the bottom and particle size progressively decreases upward.

Guyot. Any flat-topped seamount.

Gyre. Slowly rotating mass of ocean surface water.

Hadal zone. Oceanic depth zone extending below a depth of 6 kilometers.

Haline circulation. Vertical circulation of ocean water caused by variations in salinity, and therefore, density of surface water.

Halocline. Region in which salinity changes rapidly with change in depth.

Hydrographic messenger. A weight that travels down the hydrographic wire to trip the activating release of an instrument attached to the wire.

Hydrophone. A geophone used for seismic exploration at sea.

Hydrosphere. The water sphere of the earth.

Intertidal. That portion of the eulittoral zone found between high tide and low tide levels.

Invisible load. Materal transported in solution in stream water.

Island arc. Group of islands surmounting a seaward-convex submarine ridge.

Island arc trench. Trench lying on the seaward side of an island arc.

Isobath. Line connecting points on the ocean floor having equal depth beneath the water surface.

Kilometer. Metric system unit of distance equal to 0.621 miles.

Knot. Velocity of one nautical mile per hour which is equal to 0.515 meters per second (1.689 feet per second).

Lithification. Process by which sediments are transformed into sedimentary rocks.

Littoral system. Depth region that contains the eulittoral and sublittoral zones.

Marginal plateau. Gently seaward sloping surface like a continental shelf but lying at greater depth on the continental slope.

Marginal trench. A deep furrow that parallels the trend of the continental margin and lies seaward of the continental slope.

Mesoscaph. Research submersible capable of operating down to abyssal depths.

Meter. Metric system unit of distance equal to 3.281 feet.

Mid-ocean ridge. Submarine mountain range lying along the centerline of most ocean basins.

Moho. Boundary surface between the earth's crust and mantle also known as the Mohorovičić discontinuity or M-discontinuity.

Mohorovicic discontinuity. (See Moho.)

Nannoplankton. Smaller plankton and other microscopic marine life.

Nansen bottle. Device used to obtain a water sample at any desired depth.

Nekton. General term for free-swimming marine organisms.

Neritic zone. Oceanic depth zone extending from ocean surface to 0.2 kilometer depth.

Oxygen-minimum layer. Depth region in which dissolved oxygen content of ocean water is low.

Ocean basin floor. Deep ocean bottom lying between continental margin and mid-ocean ridge if one is present.

Oceanic rise. A large bulge in the abyssal floor of the ocean basin.

Paleomagnetism. Study of the relative positions of the earth's magnetic poles in past ages.

Parts per thousand. Unit used to represent salinity of ocean water.

Pelagic sediment. Ocean basin sediment derived primarily from remains of marine animals.

Phytoplankton. Plant plankton of the phyla Protozoa and Chysophycophyta.

Plankton. Single or multicellular plants or animals that drift with ocean currents.

Precision depth recorder. An extremely accurate echo sounder.

Project Mohole. United States project to drill a hole into and sample the earth's mantle.

Pteropod. Any one of a group of small marine mollusks that have wing-like lobes used for swimming.

Red clay. Ocean basin sediment composed of insoluble organic remains, terrigenous dust, and meteorite debris.

Relief. (See topographic relief.)

Reversing thermometer. Permanently recording thermometer used on a Nansen bottle.

Rift mountains. Parallel lines of mountains that flank the rift valley along the crest zone of some portions of the mid-ocean mountain range.

Rift valley. Steep-sided trough or graben lying between rift mountains.

RUM. Acronym for remote underwater manipulator.

Salinity. Measure of the weight of dissolved salts per unit volume of solvent.

Sargassum. Floating seaweed found in the Sargasso Sea.

SCUBA. Acronym for self-contained underwater breathing apparatus.

Seamount. An isolated submarine mountain rising more than 500 fathoms above the local ocean floor.

Shelf break. Line that marks the change in slope between continental shelf and continental slope.

Sial. General term for the granite-like rock of the earth's continental crust.

Siliceous oozes. Ocean basin sediment composed of the microscopic remains of marine animals that secrete a test or shell of silicon dioxide.

Sima. General term for the basalt-like rock of the earth's oceanic crust.

Sonar. Acronym for sound navigation and ranging.

Sounding lead. Weight attached to the sounding line used in bathymetric surveys.

Sublittoral zone. Depth zone lying between maximum depth of holdfast plants and greatest depth of the continental shelf.

Submarine canyon. A long, narrow, steep-walled valley cut by turbidity currents in sediments or sedimentary rocks of the continental slope and continental rise.

Suspended load. That portion of stream-transported debris that is carried in suspension.

Terrigenous sediment. Ocean basin sediment derived from weathering and erosion of continental rocks.

Test. Hard outer covering secreted by some planktonic marine animals.

Thermal circulation. Vertical circulation of ocean water caused by temperature-induced changes in surface water density.

Thermocline. Region in which temperature changes rapidly with change in depth.

Thermohaline circulation. Vertical circulation of ocean water caused by both the thermal and haline driving mechanisms.

Topographic profile. Essentially a graph of the earth's surface configuration along a given line.

Topographic relief. Difference in elevation between any two points on the earth's surface.

Trace element. Any chemical element present only in small concentration in a chemical system.

Transducer. A device used to convert energy from one form to another.

Turbidity current. Swiftly moving stream of sediment-laden water flowing in contact with the local ocean floor.

Upwelling. Periodic or continuous upward flow of water from a subsurface water mass.

Visible load. Material carried in suspension or along the bed of a stream.

Wiechart-Gutenberg discontinuity. Boundary surface between the earth's mantle and outer core.

Zooplankton. Animal plankton of the phylum Protozoa and some higher phyla.

Index